Boats, Boffins and Bowlines

Boats, Boffins and Bowlines

THE STORIES OF SAILING INVENTORS AND INNOVATIONS

GEORGE DROWER

FOREWORD BY BEN AINSLIE MBE

SUTTON PUBLISHING

First published in the United Kingdom in 2005 by
Sutton Publishing Limited • Phoenix Mill
Thrupp • Stroud • Gloucestershire • GL5 2BU

British Library Cataloguing in Publication Data
A catalogue record for this book is available from the British Library.

ISBN 0-7509-3364-X

Typeset in 11/14 pt Photina.
Typesetting and origination by
Sutton Publishing Limited.
Printed and bound in England by
J.H. Haynes & Co. Ltd, Sparkford.

Contents

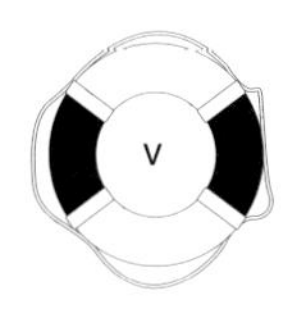

Foreword

Nothing is as wonderful as flying along in a sailing boat, whether it has the added frisson of a competitive race or is simply a relaxing Sunday afternoon sail. Having been racing for ten years, I have spent a lot of time 'messing about in boats', and winning Olympic gold with the Finn was the most glorious moment of my sailing career. Other races, such as the America's Cup, show how important small aspects of boat design can be – in some cases, the difference between winning and losing a race. That's why I was so fascinated to read George Drower's new book about sailing inventors and innovations. It gave a new insight into the history of my sport, and sailing in general. I was amused to discover that British yacht cruising didn't take off until 1830 for fear of pirates! And that King Charles II was actually of vital importance to my own sport, as he commissioned the building of the first ever English racing yacht. Signal flares, a vital safety feature of any boat, were developed by American Martha Coston, while inflatable boats were invented over a hundred years before they came into widespread use. George Drower has put together a fun and interesting book which has opened my eyes to the 'boffins' who thought of and created the nautical inventions we now take for granted. I hope it will do the same for you.

Ben Ainslie MBE
Raymarine YJA Yachtsman of the Year 2004

Introduction

Everyone has heard of the great feats of the world's circumnavigators, from pioneer Joshua Slocum, who first encircled the globe solo in 1895–8, to Dame Ellen MacArthur, who became the fastest person to complete the circumnavigation in 2005, and admired the extraordinary technologically rich racing-fit yachts from celebrity designers such as Nathaniel Herreshoff, Charles Nicholson and Olin Stephens. But what of the boating world's lesser known inventors and innovators? Their seemingly simple contraptions and ideas have been taken for granted, while they themselves have been ignored. Yet these boffins were the *real* heroes of sailing. Not only did they provide many of the basic components that made possible the creations of famous-for-being-famous designers; their ingenious devices have contributed greatly to the safety and performance of ordinary yachts.

In unearthing these forgotten pioneers, the circumstances that brought about such changes and the improvements their ideas made, this book makes some astonishing revelations. On every boat, no matter how seemingly mundane, there will be at least a handful of the simplest nautical components – from sextants to spinnakers, echo-sounders to steering wheels, signal lamps to anchors – that can provide historical stories as fascinating and rich as any associated with luxury yachts. So who were these unsung heroes? How did their ideas come about?

There was the naval engineer John Schank who, needing to devise various means of moving British warships overland to attack rebel

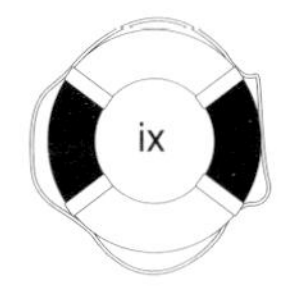

forces in lakes during the American War of Independence, thought of a method of enabling boats to manoeuvre in shallow water by equipping them with dagger-boards! When he returned to England, to serve as Dockyard Commissioner at Deptford in the 1790s, he persuaded the Admiralty to build various centreboard ships. The leaks around those dagger-board contraptions caused him to invent the watertight bulkhead; nevertheless the experimental dagger-board was abandoned in England. Its subsequent popularity in America as the centreboard obscured the fact that its originator was British.

Then there was Edward Bentall, the wealthy Essex agricultural implement maker who in 1880 audaciously applied plough technology to sailing, with a radical yacht called *Evolution*. She was the world's first boat with a fin keel, but she proved unsuccessful because Bentall had unwittingly made her design too slender. The idea was taken up and developed in Newport, Rhode Island, by Nathaniel Herreshoff in 1891, and thus the fin keel also came to be seen as an American invention.

Some winning ideas occurred through happenstance. In 1837, for example, an experimental steam launch puffing along a canal near the London Docks accidentally clipped an object in the water, which snapped off part of its propeller. Suddenly the boat went faster! This enabled a sheep farmer called Francis Smith, who was carrying out the trials, to modify the propulsion system and gain worldwide credit for inventing the world's first viable screw propeller. Then there was the story of a young Boston widow who was desperate to find a means to support her four children. In 1848 she perchanced to find, among her late husband's papers, plans to invent a signal flare. The enterprising Martha Coston went on to develop and market that safety device, but she rarely receives recognition for it. Rather unjustly the flares became synonymous with the Very pistol, because in 1877 Lieutenant Edward Very USN invented a firing mechanism for the Coston system.

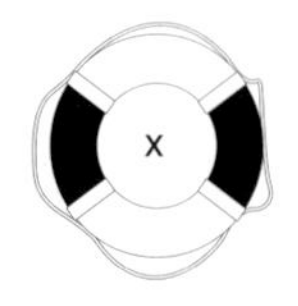

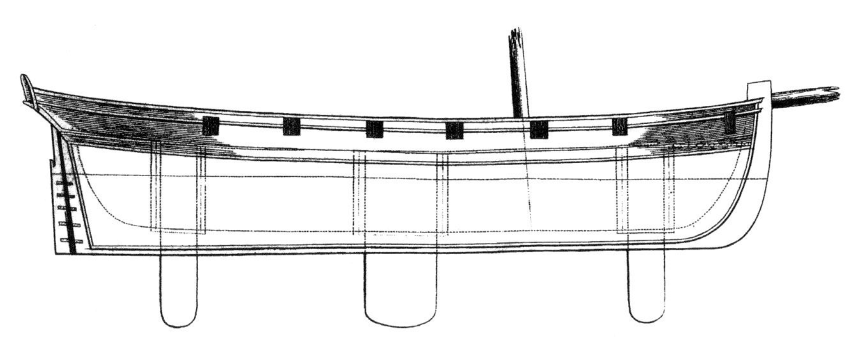

Trial, an ingenious 60ft cutter the Admiralty built to test boffin John Schank's watertight bulkhead and dagger-board inventions. (Charnock, *An History of Marine Architecture*, 1802)

Some ideas were far ahead of their time. In 1842 Peter Halkett of Richmond-upon-Thames devised the ingenious cloak-boat, a rubber dinghy sewn with the lining of a cloak! It ought to have been presented as a life-saving device but Halkett, who had tested it on rivers and at sea, was keener to promote it as an exploration boat for Arctic expeditions. Nevertheless he envisaged how giant versions of the boat could be used as emergency liferafts able to accommodate some thirty shipwreck victims. But the rapid progress of Halkett's career in the Royal Navy meant he never had another chance to develop such a system and the advantages of inflatable boats did not become widely apparent until some hundred years later. Another futuristic device was the azimuth liquid compass, perfected in 1813 by a Kent silversmith called Francis Crow. Although more stable in a boat – or yacht – than a traditional card compass, Crow's invention was unfortunately misfiled by the Admiralty as just a device for expeditions of discovery.

To be successful, even the most audacious boating pioneers needed to have an aptitude for drastic *ad hoc* innovation. Take Starling Burgess for example, the inventor of the alloy mast on the J-class yacht *Enterprise*, Harold Vanderbilt's defender of the

America's Cup in 1930. During a selection race in rough seas, *Enterprise*'s spreaders came adrift and the revolutionary 163ft mast threatened to break. On board, Burgess reacted instantly and ingeniously prevented the collapse by using spinnaker halyards to reinforce the structure – thus enabling *Enterprise* to be selected and ultimately successful in keeping the Cup for America. The world's first circumnavigating yachtsman, Joshua Slocum, faced calamity during his epic 1895–8 trip when, crossing the Indian Ocean, the tin alarm clock he used as a chronometer suddenly stopped. He cleverly restarted the timepiece by dunking it in boiling water!

Several of the boating boffins devised more than one invention, and for some the difficulty was that they were just *too* prolific. Captain George Manby made his name as the inventor of a mortar for firing rescue lines. Unfortunately he did not stop there; instead he bombarded the Admiralty with a wide variety of other schemes, ranging from a device for rescuing persons who had fallen though ice, to a message-kite, an oblong artillery shell and a harpoon gun. So when he offered them his brilliant invention of the fire extinguisher they were too irritated with him to accept it. The Revd Edward Berthon also tested the Admiralty too far with his multiplicity of schemes, which included a hydraulic speedometer (which he called a 'nautachometer'), a propeller and a clinometer. His response when they initially dismissed his scheme for a folding lifeboat was to destroy all the prototypes and move away from the sea. Only many years later was he persuaded to resume the development of the lifeboat.

Seventeenth-century boating pioneers often struggled to make progress with their schemes because they were unable to test out their ideas in advance. For the brilliant catamaran designer William Petty this meant, astonishingly, that his boats got progressively worse. Initially it all seemed to be going so well: the

20ft prototype he built in Ireland in 1662 sailed superbly well, and was said to have reached phenomenal speeds of nearly 20 knots. But his subsequent catamaran *The Experiment* perished in a storm off the coast of Spain in 1665, and an even larger multihull he created in 1684 proved unstable. All this meant catamarans were scarcely to be seen in British waters for the next three hundred years.

Meanwhile a hydrographer called Captain Greenvile Collins was also expected to accomplish more than contemporary technology allowed. From 1681 to 1687 he made the first official survey of the entire coast of Britain, equipped with nothing more than a chain and a quadrant, only to find when it was all done that British scientists condemned the primitiveness of his methodology – although mariners for the next hundred years treasured his charts for their robust simplicity.

There were a few boating pioneers whose achievements were even more influential. John Cox Stevens, who in 1844 founded the New York Yacht Club, subsequently owned the yacht *America*. That vessel's victory in a race around the Isle of Wight in 1851 laid the foundations of the America's Cup, still the premier prize in yacht racing. Other giants in the yachting world included the brilliant designer Uffa Fox, famous for his revolutionary planing dinghy and his race-winning helming abilities; and John Illingworth, with his radical yacht *Myth of Malham*. Eric Tabarly, the renowned ocean racer, led by example and inspired others to make France one of the leading yachting nations. And before them there was the intrepid Victorian adventurer John MacGregor, whose daredevil trips in canoes effectively created the sports of canoeing and canoe sailing in Europe.

Seldom was the archetypal image of an inventor more inappropriate than in the case of these boating pioneers. Such persons were far from characters of dishevelled appearance wearing

The innovative New York schooner *America*, whose winning of a cup at Cowes in 1851 started the sport of international yacht racing. (Stephens, *American Yachting*, 1904)

holey pullovers and laboriously developing their eccentric ideas in a boatshed. In fact of the pioneers covered in the fifty main stories in this book surprisingly few were proper inventors or yacht designers. Many had drawn their ideas from adjacent fields such as agriculture, engineering and especially the Royal Navy (a high proportion were captains, if not admirals). Almost invariably the boating pioneers were astonishingly important people in their own fields, and what they all had in common was a delight in and profound love of sailing.

1

Sailing Pioneers

King Charles II, Yacht Racer

The monarch who pioneered yachting in England initially found it expedient to have some first-hand knowledge of small boat sailing because of an abrupt change in his constitutional circumstances. The trouncing of his army by Oliver Cromwell at the Battle of Worcester in early September 1651 caused Prince Charles (later King Charles II) to become a fugitive, desperate to escape capture by the parliamentary forces. Determined to flee the country, he travelled incognito for six weeks until he reached the Sussex coast at Shoreham. There an unscrupulous skipper called Nicholas Tettersell, who owned a coal brig called *Surprise*, agreed to sail him over to France for 60 pieces of silver. The 34-ton *Surprise* being no bigger than a fishing boat, and with a crew of two, it is likely that Charles had to do some of the crewing himself.

On the continent the exiled prince eventually found refuge in the Netherlands, most notably at Woerden Castle near Utrecht. Charles had learnt that in Holland it had become the done thing for wealthy personages to own pleasure boats and he used the stairs leading from his rooms in the castle down to the river (which later became known as the King's Steps) to enable him to go off sailing. In such a watery country, the so-called jaghts – from a Dutch word meaning hunt or chase – which had evolved from fast fighting ships, offered

In 1651 Charles II sailed from Shoreham to France aboard this coastal collier, which he later acquired and converted into a yacht, *The Royal Escape*. (© National Maritime Museum, London)

a superb means of travel. Some of the nimbler ones were used for sport, and during the winter frosts they could even be fitted with runners and used as iceboats!

Although in fact Charles had never sailed in yachts before his exile, he had already acquired some other practical knowledge of sailing. In September 1649, during a visit to Jersey, he had learnt to sail in a frigate. It was said that he would enthusiastically take the helm for a few hours at a time and could only with difficulty be persuaded to relinquish it.

Having been restored to the throne of England in the early summer of 1660, Charles prepared to return home. On the Breda to The Hague section of his ceremonial journey his Dutch hosts supplied an escort of thirteen formal yachts. The magnificently ornate yacht that Charles himself sailed on belonged to the Board of Admiralty at Rotterdam. So impressed was Charles with the boat that he remarked to the Burgomaster of Amsterdam that he might order one of the same style immediately he arrived in England, to use on the Thames. However, the Burgomaster was so mindful of Charles's new importance that he offered to arrange for his city to acquire and present to the new king a very similar vessel that had only recently been completed in Amsterdam for the Dutch East Indies Company.

In August 1660 the yacht *Mary* duly arrived in the Thames, and at 5 o'clock in the morning she tied up at Whitehall Palace. Charles eagerly leapt aboard to inspect her. She was virtually a small Dutch warship: 52ft long, with a 19ft beam and a displacement of 100 tons, 8 guns and a crew of thirty; and with her 'wooden wing' leeboards raised she had a draught of 5ft. She had been embellished with carved and gilded ornamentation. Even Samuel Pepys assumed Charles had been impressed by these enhancements when noting in his diary the king's delight at the interior of the yacht: 'one of the finest things that I ever saw for neatness and room in so small a vessel'. But there was more to Charles than his fun-loving image suggested; he was also shrewd and competitive. According to Pepys, by November the distinguished naval shipbuilder Peter Pett had been instructed by the king 'to make one to outdo this for the honour of his country'. In so commissioning the building of the first ever English racing yacht, Charles was establishing a precedent of yacht design requiring the attentions of the best marine architects.

At Deptford Pett constructed a yacht remarkably like the *Mary*, but with one crucial difference. This involved doing away with the bulky and cumbersome leeboards (whose purpose was to resist lateral drift), instead having a deeper, 7ft draught. Called *Katherine*, in honour of Charles's bride-to-be Catherine of Braganza, she was launched in the

A Dutch yacht of the same type as the *Mary* presented to Charles II in 1660. (Ratsey and de Fontaine, *Yacht Sails*, 1948)

following April. Not long afterwards the *Anne*, a yacht virtually identical to *Katherine* and built for Charles's brother James, Duke of York, took to the water at Woolwich. Trials had showed that *Katherine*'s fine underwater form meant she was quicker than the leeboarder *Mary*, and also went closer to windward.

The new yachts provided the royal brothers with an opportunity for some sporting rivalry, and they duly arranged a race for a prize of £100. This historic event in yachting history took place on 1 October 1661 and was chronicled by the diarist John Evelyn (effectively making him the first yachting correspondent). In the first part of the contest, from Greenwich to Gravesend, the winds were contrary and *Anne* made the best use of the ebb tide to win. It was quite a festive event. Evelyn recorded: 'There were divers noble persons on board, his Majesty sometimes steering himself.' In the interval the king's barge and kitchen-boat were in attendance to supply copious food and drink. Charles did much better on the return section of the race, squaring the series by steering *Katherine* to win the return to Greenwich. It can be assumed that the Dutch raced yachts in the Netherlands but there is no written evidence for this and thus the 1661 contest on the Thames became the first authentically recorded yacht race.

Charles had a natural aptitude for sailing and according to Pepys, 'he possessed a transcendent mastery of all maritime knowledge, and two leagues travel at sea was more pleasure to him than twenty on land'. He put this expertise in nautical matters to practical effect: he reballasted *Katherine* with 4 tons of lead musket shot to improve her performance, and ensured that she and *Anne* were rigged as cutters with gaff mainsails of a type commonly used in England, rather than with Dutch-style spritsails. An opportunity to test the sea-going qualities of the new yachts offered itself during the summer of 1662 when they were sent on a mission in the English Channel, but they were driven back by a heavy storm in which they proved themselves excellent sea-boats. Pepys recalled: 'All ends in the honour of the pleasure-boats, which had they not been very good boats could never have endured the sea as they did.'

An accomplished yachtsman, King Charles II owned twenty-eight private sailing boats. (Clark, *The History of Yachting*, 1904)

In 1661 the Dutch presented Charles with yet another yacht, the *Bezan*. Just 34ft long and of only 35 tons, she had a large mainsail with a short gaff and two headsails, which contributed to her reputation as a fast mover. As with the *Mary*, Charles ordered Peter Pett to build a boat to surpass her. The king himself helmed the new yacht in a race against the *Bezan*, skippered by the Duke of York, on the Thames – and lost. Nevertheless all these experiences were eventually put to good use in 1683 with the building of the *Fubbs*, whose name derived from an old English affectionate term meaning 'chubby'. It was apparently chosen in honour of the Duchess of Portsmouth, who at the time was

Charles's favourite mistress. Reputedly designed by the king himself, the yacht had a sumptuous state room, richly decorated with carved oak and boasting a great four-poster bed adorned with gold brocade. Although better known for her pleasure activities, *Fubbs* proved to be the fastest of all Charles's yachts, partly because of her large sail area, but more especially because of her ketch rig. Charles did not, as has sometimes been claimed, invent the ketch rig, but its use in *Fubbs* certainly helped to make it respectable.

By the time he died in 1685 Charles II had owned an astonishing twenty-eight yachts, several of which he personally raced and arranged to be handed over to the Royal Navy for general service. A boat for which he retained a particular affection was the *Surprise*, the chirpy coal brig which had carried him to freedom. After the Restoration the 30-footer was sought out and purchased from her Shoreham owner; then, having been smartened and renamed *The Royal Escape*, she was kept for occasional trips on the Thames.

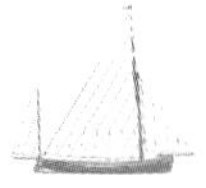

Prince Charles's voyage in the *Surprise* is commemorated each summer by the Royal Escape Race organised by the Sussex Yacht Club, www.sussexyachtclub.org.uk; and the historic 1661 contest on the Thames is commemorated with the London Frostbite Race, www.eastcoastclassics.co.uk.

John Cox Stevens and the New York Yacht Club

The most influential character in American yachting was himself from a nautical dynasty. Colonel John Stevens III, in expanding his family's merchant shipping and property development empire, had bought land in New Jersey on which came to be built the city called Hoboken. Working from his mansion at Castle Point, a rocky promontory, overlooking the Hudson River and Manhattan Island, the colonel

The First Yachting Magazine

Hunt's Yachting Magazine was the first recreational sailing journal. It was founded in 1852 by a keen yachtsman called the Hon. H.G. Hunt, who, in the year following Britain's historic defeat in the America's Cup, saw a need to facilitate discussion within the sport.

devoted his time to making experiments and inventions for the common good. He persuaded Congress to protect American inventors, and subsequently legislation was established which became the foundation of American patent law. Assisted by three of his children, in 1804 he built a propeller-driven steamboat to provide a ferry service across the Hudson to New York. Although it was revolutionary, he forfeited the claim of inventing the world's first workable screw propeller because the system could not developed as the vessel was underpowered. Inspired by the Hoboken climate of creativity, one son Robert went on to become an engineer and an ingenious naval architect, while another, Edwin, became a railway pioneer. To them is due the credit for the successful use of anthracite coal, the 'T'-rail now universally used for railways, and the revolving warship turret.

John Cox Stevens, the eldest son, was passionate about sport and though especially keen on cricket (it was reputedly he who introduced the game into America) he devoted himself to yachting. With his brothers he would build innovative models of hull forms for testing in a stream near Hoboken. In 1804, when he was just twenty-four years old, he made a 20ft dinghy called *Diver* in which he raced ferry-boats up and down New York Harbour for bets. Then in 1814 he constructed *Trouble*, a sailing canoe which was reckoned to be the first sea-going yacht in the United States. Next, in 1920, he built *Double Trouble*, an experimental but unsuccessful catamaran.

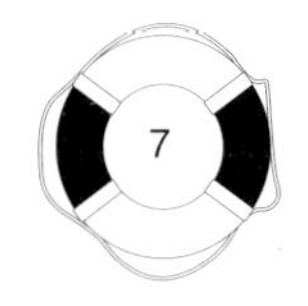

In 1844 John Cox Stevens was a principal founder of the New York Yacht Club. (© Mystic Seaport, Mystic, CT)

In 1832 he produced a 65ft waterline schooner called the *Wave*, which took part in the first yacht race in the United States. This was an impromptu event that took place in August 1835 when *Wave* rounded Cape Cod and encountered a schooner called *Sylph*, which she then raced.

It was on Stevens's brand new schooner *Gimcrack* in New York Harbour that nine wealthy yachting enthusiasts held a historic meeting on 30 July 1844. Hitherto, several attempts had been made to establish American sailing clubs, a few of which had already petered out. But the nine who gathered in *Gimcrack*'s saloon to form the prestigious New York Yacht Club (NYYC), with the keen sportsman John Cox Stevens as its Commodore, were determined to get their club off the ground. There was to be one major race each year. The first clubhouse, a purpose-built wooden structure, was sited on the Stevens's family property at Hoboken. The formation of the NYYC created an enthusiasm for competitive leisure sailing and many vessels were built. Although *Gimcrack* had been possibly the least innovative of all Steven's yachts, for a while she held her own against the newcomers, but when she was defeated Stevens decided to replace her with a new model called *Maria*, named after his wife.

Designed by his brother Robert L. Stevens, *Maria* was far ahead of her time. She had the first ever suit of sails cut horizontally, instead of the usual 'up and down' style; a streamlined hollow mast and boom; two heavily weighted iron centreboards, counterbalanced by powerful spiral springs; and outside ballast in the form of lead sheets fastened to the planking. When her bow was lengthened in a boatyard to 108 feet on the waterline she reputedly became one of the longest single-masted vessel ever built. Quickly becoming known as the fastest boat in American inshore waters, she greatly added to the growing prestige of the New York Yacht Club. However, it was an invitation contained in a letter received from England in 1850 that transformed the NYYC into the premier yacht club of the United States.

The letter was opened by a George Schuyler, one of the founders of the NYYC, and ventured to suggest that in 1851, to coincide with the Great Exhibition, a New York pilot boat be sent to Britain as an example of American shipbuilding. Stevens formed a special syndicate which agreed that, as *Maria* might be unsuitable for an ocean crossing, a new boat should be built, to give the best opportunity for the country's shipbuilding skills to be demonstrated. George Steers, the accomplished creator of the fastest pilot boats, was already known to Stevens as the designer of *Gimcrack*, and upon his recommendation the syndicate agreed to commission him to design *America*. Steers put all his expertise into the 90ft-long 170-ton schooner, which he designed with raked masts, a sharp wedge-shaped bow and a long easy form. Launched in May, *America* initially alarmed the syndicate when she was soundly beaten in trial races by *Maria*. Nevertheless, she then successfully proved her seaworthiness by crossing the Atlantic – a feat that no American racing yacht had reportedly ever achieved before – and put in at Le Havre, where Stevens and certain other members of the syndicate joined her.

On 31 July 1851 *America* anchored off Cowes, and Stevens issued a challenge on behalf of the NYYC to the Royal Yacht Squadron to race her against any number of schooners in the United Kingdom. Much to his puzzlement there were no takers. Word had already

America, owned by John Cox Stevens's syndicate, won the first ever America's Cup race in 1851. (Folkard, *The Sailing Boat*, 1901)

spread that *America* was a yacht to be reckoned with: during her voyage from Le Havre she had convincingly out-sailed a reputedly fast British yacht. Nor was there a response when Stevens pledged to stake an astonishing 10,000 guineas on another such event. Desperate for a competition, Stevens decided to enter *America* in the Royal Yacht Squadron regatta to be held for a 'Hundred Guinea Trophy' around the Isle of Wight on 22 August. Although the race was open to all foreign racing yachts, *America* was the only non-English entrant, competing against a fleet of seven schooners and eight cutters. She made history by being the first to complete the clockwise course – but only just. She had started slowly, but the fastest British yachts were stood down because of a collision, and the second place yacht at the finish, *Aurora*, might easily have won had there been a time allowance.

The trophy, which eventually became known as the America's Cup, became the property of *America*'s owners, and after her sale it was committed to the custody of Commodore Stevens, who for a while kept it at Hoboken. Some time after John Cox Stevens's death the Stevens Institute of Technology was founded on the family property at Hoboken, and for many years models of America's Cup challengers were tested in a towing tank at the institute. *America* herself changed hands many times, but was finally broken up, having been irreparably damaged in 1942 when the roof of the US Navy shed in Annapolis where she was stored collapsed under the weight of snow.

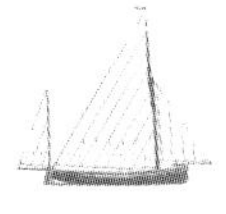

The Stevens Institute of Technology, Hoboken, www.stevens-tech.edu; New York Yacht Club, www.nyyc.org.

Richard McMullen's Corinthian Yachting

An influential pioneer of British yacht cruising, Richard McMullen was a stockbroker who lived at Greenhithe in Kent. Having ingeniously measured a revenue cutter he had inspected on the Thames at low water in the autumn of 1849, he ordered a reduced-scale version to be built, so that in 1850 he was able to take delivery of an affordable yacht. Built of pine at the commercial yard of J. Thompson at Rotherhithe, *Leo* was an 18ft, 3-ton half-decked cutter, fitted out as a miniature seagoing yacht. To her new owner she was the finest little ship in the world. Unfortunately McMullen was so keen to escape the rigours of working in the City of London that he neglected to learn the principles of sailing!

McMullen got off to a disastrous start. *Leo* sank on the day she was launched because the yard had moored her carelessly. The first time out on the river she nearly sank again when McMullen got her

The Sail Track

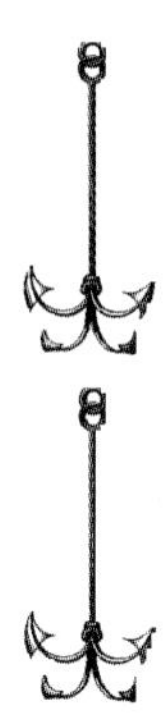

In 1840 Robert L. Stevens of Hoboken invented the sail track and slides, with which he equipped the *Onkahye*, a racing schooner of his own design. The system was intended to replace the traditional hoops by which sails were retained to a mast, but the mast track did not come into general use until later in the century.

Stevens's sail track. (Ratsey and de Fontaine, *Yacht Sails*, 1948)

masthead tangled in a brig's bowsprit. Then there was a fearful accident when a flapping sail caught him in the eye with an unmoused pair of clip hooks and almost lifted him overboard. In spite of it all McMullen pushed on, determined to become a first-rate practical yachtsman. He later wrote: 'My plan was to persevere by sail by day or night in all weathers. By getting into scrapes and getting out of them, I learnt more of practical sailing in a few months than I should have learnt in several years.' Becoming more confident as he went along, between 1850 and 1857 he succeeded in cruising over 8,000 miles in the waters between the Thames and Lands End.

Corinthian boating – the suggestion that wealthy amateur yachtsmen should play an active, hands-on role in sailing their boats – was initially articulated in the United States in the 1820s. Even so, for a long while it was reckoned there that such proletarian dalliances could only be respectably done within the confines of an institution. The New York Yacht Club had been founded with Corinthian ideals in 1844, while the membership of the Seawanhaka Corinthian Yacht Club (formed in 1871) was reserved exclusively for owner-sailors. The term spread to Britain in 1872 with the formation of the Royal Corinthian Yacht Club. In practical application it meant that gentlemen might steer during races, although cruising yachts were still assumed to need strong and seasoned professional crews. In Britain, however, cruising was still not commonplace, for few yachts had dared venture far from the coast for dread of pirates lurking in the English Channel. It was not until 1830, when a French fleet destroyed numerous pirates' vessels while bombarding Algiers into submission, that the threat of the marauders diminished.

At the time when McMullen was learning by his mistakes in the 3-ton *Leo*, yachting was regarded as something of a social ceremony at which the presence of spectators was a necessary adjunct. Unimpressed by the shenanigans of 'Cowes yachting', McMullen took delight in striving to perform every nautical task, regardless of how trivial, to as high a standard as possible. Although a self-taught amateur, he was uniquely professional in his actions – maintaining his boat to near naval standards of cleanliness and efficiency, and keeping his equipment in good order. He went on to raise the standard of seamanship in amateur sailing to a far higher level than had been reached before by the publication in 1869 of his book *Down Channel* – a work which effectively established him as Britain's first cruising Corinthian yachtsman. An endearingly honest account of the daft muddles into which he got himself in his early boating years, *Down Channel* progressively showed how he had ingeniously used the opportunity of adverse circumstances to evolve new skills, which he could then apply to bigger boats.

In 1858 *Leo* was replaced with a 32ft cutter called *Sirius*, which also derived from the original specification he had obtained in 1849. Sometimes he sailed her with a paid hand, and later with his wife, and in all *Sirius* covered some 12,000 miles, which included trips to the Scilly Isles and Ireland, and a circumnavigation of Great Britain. In 1865 McMullen built an even larger version, the 42ft lugger *Orion*, which required several hands to crew her. In his writings

The 3-ton cutter in which Richard McMullen, the pioneer of Corinthian cruising, sailed thousands of miles in the 1850s. (Heaton, *Yachting*, 1955)

McMullen disclosed how in Cherbourg he had sacked *Orion*'s crew for being insolent and smoking too much! Undaunted by the challenges of sailing the 19-ton vessel home alone, he devised techniques to do so. One involved reefing the mainsail before the onset of bad weather; it took him two hours to do, but he accomplished it.

His book and other sage advice, simply based on his years of personal experience, won him several important emulators: literary yachtsmen such as Claud Worth, Frank Cowper, Frank Knight and Francis Cooke (who became so enthusiastic that he later wrote *The Corinthian Yachtsman's Handbook*). Their own short-handed cruises in converted working boats generated wider interest, and led to the creation of institutionalised Corinthian-minded cruising organisations such as the Royal Cruising Club (1880) and the Cruising Association (1908).

The Boat Show

The Royal Aquarium, Westminster. (*Westminster City Archives*)

The first recreational boat show was the 1894 International Yachting Exhibition at the Royal Aquarium, Westminster. That February, reported the new magazine *Yachting World*, the public could inspect the very latest yacht designs and chandlery for the admission price of a shilling. The Royal Aquarium itself was an entertainment hall (demolished in 1904) on the edge of Parliament Square, with winter and summer gardens, a swimming pool and huge tanks containing fresh and saltwater fish. It was also well known for hosting chrysanthemum and billiard shows, and circuses at which ladies were shot from cannons.

Most distinctively, McMullen was the first yachting writer to establish what yachts ought to do in stormy weather. In *Down Channel* he wrote some of the wisest words about what to do in such difficult conditions: 'I am convinced that unless a small vessel, especially an open one, can be got into harbour before the sea

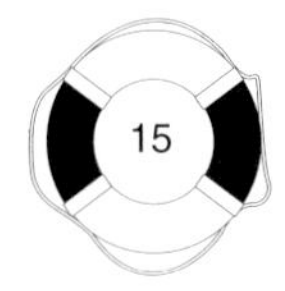

becomes very heavy, there is more safety in keeping the deep waters and in not attempting to approach the land at all.'

McMullen himself, as the years passed, became slightly eccentric. A staunch Protestant, he produced pamphlets attacking Catholicism, and even peppered the two otherwise excellent books he produced after *Down Channel*, *Orion: How I Came to Sail Alone in a 19-ton Yacht* (1878) and *An Experimental Cruise Single-Handed in the 'Procyon'* (1880), with ravings against screw-steamers, mechanically propelled vehicles, compulsory education, trade unionism and yet more Popery.

His death was in the Viking tradition. French fishermen found him dead on 15 June 1891, sitting at the helm of his 27ft lugsail yawl *Perseus* in the English Channel. An autopsy found he had suffered a heart attack 24 hours before.

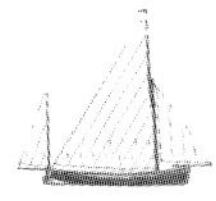

The Royal Corinthian Yacht Club, www.royalcorinthian.co.uk, the Royal Cruising Club, www.rcc.org.uk, and the Cruising Association, www.cruising.org.uk.

James Gordon Bennett's Transatlantic Race

A leading participant in the first transatlantic race was the heir to a newspaper publishing fortune created by his father James Gordon Bennett, who pioneered many of the techniques of modern journalism. A workaholic Scotsman, Bennett had started the *New York Herald*, writing the entire newspaper himself and making up for the lack of news by producing sensational opinions, fictitious intelligence and reckless personal attacks.

His son James Gordon Bennett junior was brought up in the same exclusive world as other Rhode Island elite families, such as the Du Ponts and the Vanderbilts, who made ostentatious displays of their wealth. Rogue though master James was, he seems to have escaped becoming a total wastrel through his aptitude for sailing, which became apparent in 1857 when for his sixteenth birthday he received from his father a 77-ton centreboard sloop named *Rebecca*. He first appeared on the yachting scene that summer in a series of New York Yacht Club races. Although *Rebecca* did not win any of the events, the precocious skipper must have handled himself and his crew of twenty-one well enough, because when all the captains convened he was duly elected to membership. He remains the youngest member ever admitted, at 16 years and 3 months. The next year, in a race from Staten Island around Long Island to Throgs Neck, *Rebecca* won by five hours but was disqualified when other members accused Bennett of taking a short cut through Plum Gut. The performance showed his predilection for sharp practice, and also bold seamanship, Plum Gut having been deliberately omitted from the course because of the risks involved.

In 1861 Bennett, now aged twenty, progressed to the ownership of the new 205-ton racing schooner *Henrietta*, designed by Henry Steers. Later that year James Bennett senior, having resolved that the *Herald* would support the North in the American Civil War, arranged with the Lincoln administration to lend *Henrietta* to the Revenue Cutter Service. Young Bennett was given the rank of lieutenant and served as the vessel's commanding officer for a year, during which time she patrolled off Long Island and then went to Florida to participate in the capture of Fernandina. Bennett became by necessity an accomplished sailor and navigator, learning skills that he was to exercise for the rest of his life.

The historic event which enabled him to put that know-how to good effect stemmed from a conversation at the Union Club on 26 October 1866 between the tobacco tycoon Pierre Lorillard junior and Franklin Osgood, whose father-in-law was Cornelius Vanderbilt. They were boasting about their respective vessels – Osgood's deep keel *Fleetwing* and Lorillard's new centreboarder *Vesta* – and

James Gordon Bennett, the wealthy yachtsman whose *New York Herald* funded daredevil sporting events. (Rayner and Wykes, *The Great Yacht Race*, 1966)

comparing their merits, and under the influence of alcohol the discussion spiralled out of control into a wager for $30,000 each to race across the Atlantic. On learning of this proposed escapade, Bennett eagerly joined in. Quite unwittingly, the *Herald* had played a part in the audacious choice of race route when, several months earlier, it had urged America's 'smooth water gentry' to 'trip anchors and start out on a cruise on blue water'. But it would be risky. Hitherto America's finest racing yachts had mostly sailed in sheltered waters, and furthermore the winter weather was closing in. As they sobered up Osgood and Lorillard wisely decided to remain on land and not accompany their crews, but Bennett was determined to go.

Bennett prepared *Henrietta* well, reducing her towering rig, replacing her tiller with wheel steering and building a heavy shelter over her cockpit. He also hired to assist him Samuel 'Bully' Samuels, a harsh taskmaster who already held the transatlantic speed record in the clipper *Dreadnought*. Sensing disaster, *Henrietta*'s crew deserted her just before the start of the race on 11 December 1866, so they could only be replaced with inexperienced personnel. The yachts were similar in specification, so for the first week out from Sandy Hook they remained in sight of each other until a mighty storm on 18 December. A huge breaker staved in *Henrietta*'s seaboat and opened up her deck seams, and for a while she was obliged to heave-to until the weather eased. Tragedy then struck *Fleetwing* when a freak wave plucked six crewmembers from her exposed cockpit. All perished! High winds prevailed all the way across the Atlantic. There was no beating to windward, and the race was an unremitting drive. Bennett arrived at Cowes on Christmas Day, *Henrietta* having made the trip in the amazingly quick time of 13 days, 21 hours and 55 minutes. *Fleetwing* and *Vesta* showed up on Boxing Day, only 40 minutes apart. As the first yacht race across the Atlantic the contest aroused sensational interest, much of it fanned by the *Herald*'s reporters who used the new transatlantic cable to send race reports back home. Bennett's victory in what was dubbed 'The Great Ocean Race' had been emphatic. As the only owner brave enough to be aboard, he returned home an American hero.

In 1867 Bennett took over his father's newspaper publishing business, and set about raising the already successful *Herald*'s circulation with fictitious stories (such as a hoax that wild zoo animals were rampaging in Central Park) and by sponsoring special expeditions that would generate popular concern. One famous example was the sending of Henry Morton Stanley to find the lost Dr Livingstone; he also funded a US Navy expedition to the Arctic. Tragically most of its members perished, but that too created news and sold papers. Such escapades were well within Bennett's means: his guaranteed yearly disposable income was estimated to be a million dollars. But on New Years Day 1877 he overstepped the mark. Having drunk too much at a party given by his fiancée's parents, he urinated in a fireplace. The resulting disgrace caused him to fight a duel, before withdrawing from New York society.

Eventually settling near Paris, on Louis XIV's estate at Versailles, he continued to administer the *Herald* via telegraph communications, and established the *International Herald Tribune* which reflected his maritime interests by including storm warnings

Bennett won the1866 transatlantic race in his yacht *Henrietta*. In 1870 his schooner *Dauntless* (left) raced westwards against the America's Cup challenger *Cambria*. (Heaton, *Yachting*, 1955)

transmitted from America. To further publicise his newspapers he 'created' news events by sponsoring various sporting competitions for Gordon Bennett trophies in motor racing, power boating, aeroplane races and ballooning. He also directed his business from his magnificent steam yachts, the beautiful 226ft *Namouna* (built in 1883) and the 301ft *Lysistrata* (1900), which even had its own motor car and private suites to accommodate his travelling harem of female admirers. Although these lavish yachts had crews of nearly one hundred, whenever he was at sea Bennett insisted on doing all the navigation himself.

The New York Yacht Club, www.nyyc.org; and the *International Herald Tribune*, www.iht.com.

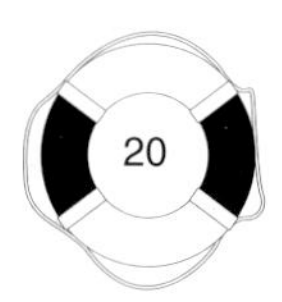

EASTBOUND TRANSATLANTIC RECORDS

TYPE OF BOAT	NAME	TIME	SKIPPER	YEAR
Motor-ship	*Cat-Link V*	2d 20h 9m	Claus Kristensen	2003
Catamaran	*PlayStation*	4d 17h 28m	Steve Fossett	2001
Monohull	*Mari Cha IV*	6d 17h 53m	Robert Miller	2003
Trimaran	*Primagaz*	7d 2h 34m	Laurent Bourgnon	1994
Rowing dory	*Skoll 1080*	55d	Tom McClean	1987

John MacGregor, Boating Celebrity

In 1866 there appeared a sensationally bestselling book called *A Thousand Miles in the 'Rob Roy' Canoe*, a vivid account of an audacious sailing and paddling trip written by an adventurer who, in effect, invented for Europeans the sport of canoeing. In fact John MacGregor's maritime adventures began when he was just five weeks old. He was sailing to India with his parents when the merchant ship they were on caught fire; they would doubtless all have perished had not a passing brig sighted the burning vessel and rescued the survivors. MacGregor's father was General Sir Duncan MacGregor, whose military postings meant John frequently had to change schools, but this seemed to make him versatile. John became passionate about various adventurous sports, and especially boating. Seizing every chance to get afloat, when he was aged twelve he even helped to crew a lifeboat. His religious faith grew so strong that for a while he considered becoming a missionary. Also fast developing were his remarkable talents for mechanical engineering, drawing and writing; and while still a student he began to have articles published in the *Mechanics' Magazine* and *Punch*. But as a lad from a privileged background, he had already established the practice of giving to charity all that he earned from such activities.

Awarded a degree in mathematics while a student in Ireland in 1840, MacGregor went on to qualify as a barrister and then specialised for a time in patent law. But despite being – as the *Dictionary of National Biography* later described him – possessed of ample means, he threw aside the chances of a good law practice, choosing instead to devote the rest of his life chiefly to active philanthropic work and foreign travel. In July 1849 he started an overland journey across Europe to the Levant, Egypt and Palestine, and in 1851 he went to Russia, and then on to Algeria, Canada and the United States. During this time he learned mountaineering skills, climbing Mont Blanc, Vesuvius and Etna one after another, and duly wrote to *The Times* to let the world know what he had done. Always he illustrated his diaries with his own sketches, such as those published in 1859 of his earlier tour of North America, and in 1857 he had even produced the illustrations for Dr David Livingstone's *Missionary Travels and Researches in South Africa*. Legend has it that in the spring of 1865, having somehow injured his hand so that he was unable to take part in a important shooting competition in Wimbledon for the London Scottish Volunteers, in which he was a captain, he inexplicably decided instead to try canoeing. He seems to have thought that a voyage in such a small boat might be an ideal means by which many of his talents could be put to good use for substantial charitable fund-raising.

Years earlier he had tried animal-skin Kamchatka canoes in Russia and birch canoes in New Brunswick, but now MacGregor set about designing a quite evolutionary wooden 'decked' canoe which became known as a kayak. Built for him by Searle & Sons of Lambeth, the 80lb cedar and oak boat was 15ft long (so he could if necessary squeeze it into a German railway wagon), just 28 inches wide and 9 inches deep, and it was equipped with a 7ft double-bladed paddle, plus a lugsail and jib. He christened her *Rob Roy*, which was his parents' nickname for him. The intended journey

John MacGregor, the adventurous 6ft 6in barrister who developed the sport of canoeing. (Trinity College, Cambridge)

across a thousand miles of rivers and lakes in continental Europe had never before been attempted in such a tiny craft, and nor were there available many reliable maps of the route (even the Paris Boat Club was said to know nothing of French rivers).

From Westminster on 9 July 1865 MacGregor paddled downstream to Greenwich where, he later wrote, he hoisted the mainsail: 'A fine breeze filled the new sail, and we skimmed along with a cheery hissing sound.' He took the Channel steamer to Ostend. Equipped with only a spirit stove, a fork and spoon carved from one piece of wood, and 9lb of luggage, his great European adventure began. He spent three months on the rivers and canals of Belgium, France and Germany. He also ventured on to Lake Constance and various Swiss lakes, and descended the rapids of Bremgarten on the hazardous Reuss.

Capitalising on his natural genius for publicity, MacGregor had put together a deal with the *Record* newspaper to print week by week his account of the tour as he went along. The novelty of his mode of travel and the vividness of the reports he filed meant that he received newspaper coverage not only in Europe but also in the United States, Canada and South America. Sometimes when *Rob Roy* stopped overnight hundreds, and occasionally thousands, of onlookers would gather in the mornings to watch the towering 6ft 6in canoeist set off. In many respects the trip was one of the first boating 'media events', attracting interest across several continents. When he finally paddled under Westminster Bridge on his return on 7 October 1865 he was the hero of the hour.

Within weeks MacGregor's account of the adventure was published, complete with illustrations taken from woodcuts he had done himself, and it became the top non-fiction bestseller of 1866.

On his expeditions MacGregor would sometimes be applauded by hundreds of spectators. (MacGregor, *A Thousand Miles in the 'Rob Roy' Canoe*, 1883)

It was of historical significance too, foreshadowing with its terse, fast-paced prose a shift in style towards that of modern travel books. In addition to donating all the profits derived from the book to the Shipwrecked Mariners' Society and the National Lifeboat Institution (later RNLI), the philanthropic MacGregor embarked on a nationwide lecture tour to raise more money for charities. A natural showman, he would appear on innumerable town hall platforms with his canoe, paddle, sails and various other props, then he would slip behind a screen and emerge in the eccentric Norfolk jacket uniform he had devised for canoeing.

The formula for success thus established, in the summer of 1866 MacGregor took a break from his public speaking activities to build a new *Rob Roy* canoe and then set off for the Baltic. Reaching Norway by steamer, he explored lakes and rivers, ventured into the Baltic near Stockholm, crossed the narrows to Denmark, paddled through Schleswig and, via the North Sea, eventually returned to London. Having recounted his experiences in a new book and completed another fund-raising tour, in 1867 he was off again, this time to Paris where Emperor Napoleon III had invited him to the Paris Exhibition. To sail there MacGregor built at Cowes a 3-ton slender 21ft yawl which he described in another travelogue, entitled *The Voyage Alone in the Yawl 'Rob Roy'*. The last *Rob Roy* sailing canoe, made in 1868, had several ingenious features, including a topmast made from a section of a fishing rod, deep blue sails to temper the glare of the sun, and a removable aft deck to create a sleeping space under a mosquito net and awning. With this boat he made the first transit by a leisure craft of the Suez Canal (which was still under construction), paddled down the Red Sea and ultimately reached Jordan.

The *Rob Roy* books were reprinted many times, and at least one edition had an Introduction by Arthur Ransome. As well as being exciting travel yarns, they were also full of practical hints, such as how to steer around obstacles in fast-flowing streams. The books

Rob Roy, the sailing canoe MacGregor famously took across Europe in 1865. (MacGregor, *A Thousand Miles in the 'Rob Roy' Canoe*, 1883)

inspired many would-be boaters to make long leisure trips on inland waterways, as did the Royal Canoe Club, founded by MacGregor in 1866. With this, in effect, he created the sports of canoeing and canoe sailing.

MacGregor's original *Rob Roy* canoe is preserved at the National Maritime Museum, Greenwich, www.nmm.ac.uk; there are historic canoes at the River and Rowing Museum, Henley, www.rrm.co.uk; also visit the Royal Canoe Club website, www.royalcanoeclub.com.

Joshua Slocum's Circumnavigation

One morning in April 1895 a yacht slipped out of Boston harbour and headed off into the Atlantic on the first leg of an unprecedented solo attempt to sail around the world. The skipper of this daredevil

venture was 51-year-old Joshua Slocum, a gruff but bookish and multi-talented sea captain. Born in Nova Scotia, Canada, he had earned his various tickets through diligent self-education and by virtue of a strong practical mind, and by 1887 he found himself the owner-skipper of a large trading barque in South America. Financially ruined when the uninsured ship was wrecked on the Brazilian coast, he resourcefully salvaged one of her boats, which he rigged as a yacht and sailed home to Boston with his wife and four children – a journey of some 5,500 miles. Needing to earn some money he even wrote a book, *The Voyage of the Liberdade*, describing the trip, but although it was well reviewed it sold poorly. Worse still, as sailing ships became increasingly superfluous in the age of steam, Captain Slocum's prospects of returning to sea were bleak. He was reduced to doing casual jobs along the waterfront as a carpenter and rigger.

A glimmer of hope came in March 1892. Learning of Slocum's plight, a retired Massachusetts mariner told him: 'Go to Fairhaven and I will give you a ship'. Then he added mischievously, 'But she needs some repairs'. The next day Slocum landed at Fairhaven to discover that he had been the victim of a humorous practical joke. The 'ship' proved to be a 37ft Delaware oyster sloop called *Spray*, rotting away in a field and covered with canvas. Undeterred, Slocum immediately accepted ownership of the hundred-year-old vessel, reckoning that, properly restored, she might just enable him to earn a living from coastal fishery. Unperturbed by *Spray*'s hopelessly dilapidated condition, he laboured for thirteen months in all weathers, substantially rebuilding her. He industriously axed a nearby oak tree for a replacement keel and timbers, mixed concrete for ballast and even made bolts for fastenings.

No sooner was *Spray* launched in 1894 than Slocum realised he lacked the mindset to catch fish commercially. But happily he soon discovered that the 13-ton craft had several very special qualities:

The yacht Captain Slocum successfully rebuilt before circumnavigating the world. (Slocum, *Sailing Alone Around the World*, 1900)

she was exceptionally fast and had an astonishing talent for steering herself. Exactly what gave her that directional stability Slocum could never really fathom. In the rebuilding he had not altered her underwater profile, created by her waterline length of 32ft, generous beam of 14ft and draught of just 4ft. *Spray*'s excellent seakeeping qualities, and the absence of any prospect of mercantile work, led Slocum to wonder if he might not achieve fame and fortune by sailing her around the world.

Single-handed voyages even across the Atlantic were still comparativelt rare. As yet, the only circumnavigation made by a yacht had taken place in 1876–7, when the *Sunbeam*, a lavishly equipped 531-ton schooner belonging to a British millionaire, had completed the trip with a crew of thirty-two including a lady's maid and a butler. In contrast the impoverished Slocum, unable to afford even a chronometer, would need to navigate using a corroded tin alarm clock bought in a junk shop for a dollar, sounding lead and a rotating log, and would have to deduce longitudes by lunar distance. He would also rely on his gung-ho but extraordinarily accurate innate ability as a dead-reckoning navigator.

Having crossed the Atlantic via the Azores, Slocum arrived at Gibraltar. Heeding the warnings from British authorities of dangerous pirates in the Red Sea, he abandoned his plans to continue through the Mediterranean and instead opted to try to find a westabout route around the world. To ready *Spray* for the stormy waters of the southern oceans, when he reached Brazil he used his improvisational skills to alter her rig from sloop to yawl; then he made his way down the South American coast, battled for three weeks through the Straits of Magellan and then island-hopped across the Pacific to Australia. A phenomenally trustworthy self-steerer, *Spray* excelled herself during the 23-day run through the Coral Sea, throughout which Slocum's total time at the wheel was just 3 hours. Disaster seemed imminent in the Indian Ocean when the second hand of the tin clock fell off and the timepiece – the only one he had on board – stopped. But ingeniously he got it going again by dunking it in boiling water! Rounding the Cape he crossed the Atlantic via St Helena and Ascension Island, where the milk goat he carried on board munched through his charts. Finally, having narrowly dodged a minefield off New York, on 27 June 1898 he completed his voyage when he reached Newport, Rhode Island.

Slocum had accomplished the epic 46,000-mile voyage in three years and two months and made maritime history as the first person to circumnavigate the world single-handed (and some of the time *against* the prevailing winds). It was an astonishing achievement, which effectively made him the first celebrity yachtsman. On the lecture circuit Captain Slocum was more popular than the raconteur Mark Twain, and invitations flooded in for *Spray* to star in international exhibitions, while the publication

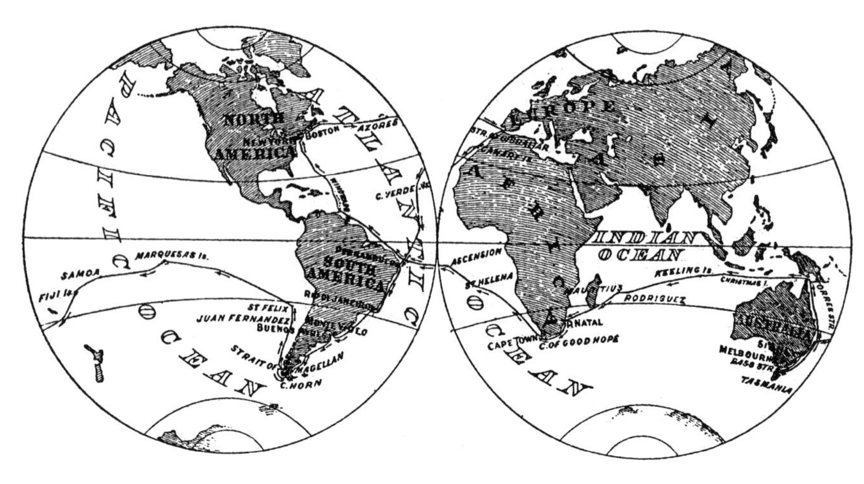

Slocum's epic route, 1895–8. (Slocum, *Sailing Alone Around the World*, 1900)

in 1900 of his account of the voyage, *Sailing Alone Around the World,* sent Slocum to the top of the bestseller list (it has subsequently never been out of print). One of the finest sea stories ever, the book also narrated with wry literary humour his adventures in the thirty-four various places he stopped at during the trip. With endearing modesty he trod lightly over the severe dangers he had faced at sea, but used his book to freely impart details of the special sailing techniques he had used, and indeed he encouraged other yachtsmen to embark on similar trips. His achievements provided both inspiration and motivation to future generations of daring round-the-world yachtsmen, such as John Voss (1901–4), Harry Pidgeon (1921–5), Alain Gerbault (1924–9) and then Francis Chichester (1966–7). *Spray* too was an inspiration. After 1909, when the popular American yachting magazine *Rudder* published her plans, many hundreds of replicas were made of her in wood, steel and eventually even GPR in several countries.

The First Yachtswoman

Annie Brassey, the first yachtswoman to circumnavigate the world, and, below, the *Sunbeam*. (Brassey, *A Voyage in the 'Sunbeam'*, 1880)

In 1877 Lady Annie Brassey became the first yachtswoman to circumnavigate the world. Her millionaire husband's three-masted topsail schooner, the luxurious 157ft *Sunbeam*, had a crew of thirty-two, including a butler, two maids and a nanny for her children. Subsequently her vivid account of the ten-month adventure, *The Voyage in the 'Sunbeam'* (1878), established Annie Brassey as the first female best-selling yachting writer.

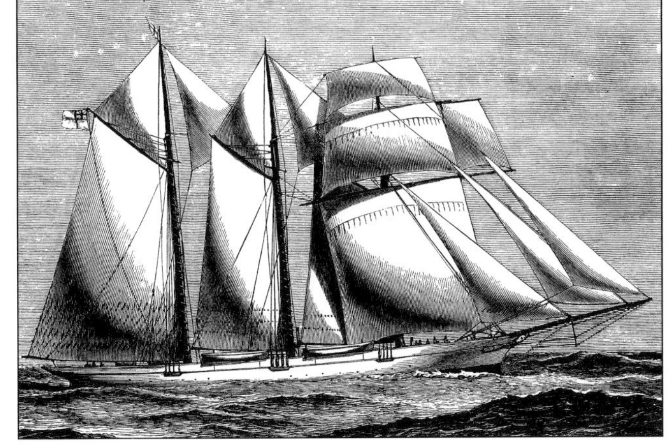

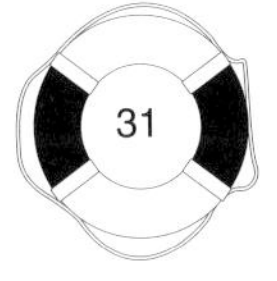

Slocum's many skills as a navigator, helmsman and shipwright enabled him to succeed, but as a maritime pioneer he had also been lucky. After his famous voyage he somehow never really settled down ashore. Notwithstanding the security of his new wealth, he was unable to resist pushing his luck again, and took *Spray* back to sea as a trader (although such activity was now uneconomic). In 1909 he left New York for Latin America, but tragically disappeared without trace. Since no storm was reported in the area, it came to be believed that *Spray* had been run down, ironically probably by a steamer, while her skipper was asleep below.

The Joshua Slocum Society International website: www.joshuaslocumsocietyintl.org.

Eric Tabarly, the Yachting Populariser

The yachtsman whose exploits were largely responsible for transforming France into a top sailing nation initially came to prominence in the 1964 *Observer* Single-Handed Transatlantic Race (OSTAR). Lieutenant Eric Tabarly had until then been an unknown 32-year-old pilot in Lancasters with the Aeronavale in Indo-China. But having obtained leave from the French Navy to participate in the race, he devised a lightweight monohull which he built of plywood to reduce both costs and weight. At 44ft, *Pen Duick II* was far longer than was generally thought suitable for single-handed sailing, so to ensure that the sails were manageable Tabarly opted for a ketch rig. Innovatively he also fitted her with several special features including a car steering wheel and an internal helming position with an aircraft-type Perspex dome in the coachroof. *Pen Duick II* went to the starting line at Plymouth as the only yacht designed specifically to win the race.

One of sixteen entrants, Tabarly was up against Francis Chichester, who had won the initial OSTAR four years earlier. It was a remarkable performance from the Frenchman. Forced to sail the final 2,000 miles of his first transatlantic race without self-steering after *Pen Duick II*'s wind vane broke, Tabarly could not leave the helm. For the last two weeks of the voyage he could only catnap and he got no sleep at all during the final 48 hours. Throughout the voyage Tabarly maintained radio silence with the outside world. Not only did he surprise everyone by winning, he also easily broke the thirty-day barrier for the 3,000-mile crossing, beating Chichester's record by thirteen days. His victory made such an impression in France that President de Gaulle immediately awarded him the Légion d'Honneur, and Tabarly flew home to a tickertape procession down the Champs-Elysées. Not since 1929, when Alain Gerbault completed his four-year circumnavigation of the world in the cutter *Firecrest*, had there been a French national sailing hero.

The original *Pen Duick* (meaning 'coal tit' in Breton) was an elegant 49ft gaff cutter built in 1898 to a design by William Fife III, and purchased by Eric Tabarly's father in 1938. It was in this physically demanding boat in the waters around Brittany that Eric discovered his talent for sailing. Laid up and left at the mercy of the elements during the war, by 1947 the boat had deteriorated to such an extent that she was quite unsellable, and so she was given to Eric to restore. But by 1958, when the young pilot returned from Saigon with enough money to begin the restoration, she had virtually rotted beyond repair. Undeterred, and using the hull as a mould, Tabarly built a new fibreglass hull which he then fitted out using as many of the original yacht's fittings as possible. It was a pioneering approach, and when relaunched in 1959 *Pen Duick II* was for a while reputedly the longest fibreglass yacht in the world.

Tabarly's innovative and uncompromising approach next became apparent in 1967 with *Pen Duick III*, a wishbone schooner. Although the new measurement rules insisted that yachts should have an engine, there was no requirement for the engine to work, so Tabarly installed a non-functioning engine upside-down to lower the boat's centre of gravity. That year it helped *Pen Duick III* to victory in the Fastnet, Sydney–Hobart and Channel Races. The next monohull, the 74ft maxi *Pen Duick VI*, was fitted with a spent uranium keel. For her unsuccessful campaign in the 1973–4 Whitbread Round the World Race she had required a crew of fourteen, so when Tabarly entered the huge ketch for the 1976 OSTAR it seemed she would be too much for a single crewman to handle. The race proved to be one of the toughest ever, as severe weather pounded the fleet, and Tabarly's self-steering gear was shattered only a thousand miles out in the Atlantic. But he pressed on, steering by hand and maintaining almost complete radio silence, and again astonished everyone when *Pen Duick VI* slipped quietly into Newport at night after only twenty-three days at sea. Tabarly was the first yachtsman to win the event twice; there were more wild celebrations in Paris, while Tabarly was awarded an even higher honour, becoming an Officier de la Légion d'Honnour, and attained legendary status in the sailing world.

Tabarly next turned his attention to multihulls. A test sail in 1968 on the 42ft British trimaran *Toria*, which had won the 1966 Round Britain and Ireland Race in record time, tempted Tabarly to build a multihull of his own. The result was an astonishing wing-masted ketch, the 67ft aluminium trimaran *Pen Duick IV*, weighing just 6.5 tons. Nicknamed the 'Giant Octopus', she was rushed to the start of the 1968 OSTAR, though she was not quite ready and soon had to quit the race because of mechanical breakdowns and a collision. According to an extensive obituary of him in *The Times*, Tabarly went on to champion the cause of multihulls, becoming one of the first to develop hydrofoils for large multihulls, and building some of the most radically innovative boats. Particularly effective was his 46ft foiler trimaran *Paul Ricard*, which in 1988, in a voyage lasting just 10 days and 5 hours, smashed the record for the New York–Lizard transatlantic crossing famously set by the schooner *Atlantic* as long ago as 1905.

Tabarly's high-tech *Pen Duick IV* inspired a generation of French multihull racers, and – skippered by Alain Colas – won the 1972 Single-handed Transatlantic Race. (Beken)

A crucial factor in Tabarly's rise to superstardom had been the French Navy's funding of his boats. After his victory in 1964 he was seconded to the French Ministry of Youth and Sport, and then was made Inspector of Sailing at the Fontainebleau School of Interservice Sport. Quite at variance with his gruff monosyllabic conversational style, he became a prolific author of sailing books (only a few of which were ever published in English). Reciprocating the government's support, he taught the national servicemen who helped crew his yachts all he knew. So inspirational was he that many of his assistants, such as Jean-François Coste, Marc Pajot, Philippe Poupon and Titouan Lamazou, went on to become France's leading yachtsmen, as recognisable and famous as leading French footballers. Some occasionally even bettered Tabarly. One protégé, Alain Colas, won the 1972 OSTAR in the trimaran *Pen Duick IV* – which Tabarly had needed to sell for tax reasons – and then in 1973–4 sailed around the world in a record 168 days. So inspirational were Tabarly's accomplishments that special events such as the Route du Rhum (since 1978) were established to cope with the mushrooming of popular enthusiasm for short-handed long-distance yacht races. A *Yachting World* obituary of Tabarly, noting how yachting had grown into a national sport in

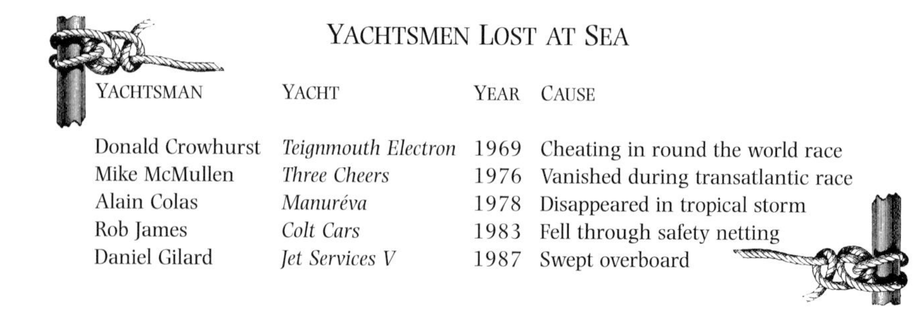

Yachtsmen Lost at Sea

Yachtsman	Yacht	Year	Cause
Donald Crowhurst	*Teignmouth Electron*	1969	Cheating in round the world race
Mike McMullen	*Three Cheers*	1976	Vanished during transatlantic race
Alain Colas	*Manuréva*	1978	Disappeared in tropical storm
Rob James	*Colt Cars*	1983	Fell through safety netting
Daniel Gilard	*Jet Services V*	1987	Swept overboard

France, second only in popularity to cycling, stated: 'In little more than ten years the feats of this one man had inspired a generation of sailors, and sailing had become the nation's second biggest sport.' The growing popularity of sailing enabled mass-production French boatbuilders to reduce their unit costs and widen their product ranges, so that yachts by Beneteau, Dufour and Jeanneau even became commonplace in British marinas.

But for Tabarly it all ended tragically in 1998. Virtually retired from competitive sailing, he was taking his family's beloved boat, the original *Pen Duick*, to a classic boat rally in Scotland. En route, off the Welsh coast on the stormy night of 13 June he was accidentally knocked overboard and drowned. Having established a brilliant reputation for exemplary seamanship in the world's toughest high-tech ocean racers, it was ironic that he should have been so careless as to perish for wont of a simple illuminated lifejacket.

The most notable French websites on Eric Tabarly are www.penduick.com and www.asso-eric-tabarly.com.

2

Navigation and Pilotage

Greenvile Collins's Coastal Charts

In 1667 a fleet of Dutch warships led by Admiral Michael de Ruyter raided the Thames estuary, destroyed warships in the Medway and advanced as far as Chatham. The raid was a national humiliation, and it indicated that the Dutch had a detailed knowledge of the English coast. Since no official survey had ever been made of Britain's coast, English mariners had to rely on a hotchpotch of rough charts called 'sea cards'. Produced by a group of cartographers known as the 'Thames School', even the sea cards were derived from Dutch charts. When one of their number, John Seller, published *The English Pilot* in 1670 his pretended new book was compiled mainly from Dutch plates, and even its newly engraved plates were copied from Dutch originals. Then the Dutch began copying Seller's plates in order to sell them to the English, knowing they would find a good market! By 1681 Charles II, having consulted with Samuel Pepys, the Secretary of the Navy, decided something really must be done, and in June issued a proclamation announcing that Captain Greenvile Collins had been appointed 'to make a survey of the seacoast of the Kingdom'.

An experienced officer with the Royal Navy, Collins had initially shown promise as a mapmaker while serving on Sir John Narborough's 1669–71 voyage of exploration, sponsored by the

Duke of York, to the Straits of Magellan and along the Patagonian coast. The remarkable charts that Collins compiled of the Magellan Strait were generally used for many years by all expeditions entering the Pacific. Subsequently, as sailing master of Narborough's flagship *Plymouth*, he was not only responsible for navigation but also had to prepare drafts of harbours on uncharted coasts, whose features he also noted with hydrographic sketches. In 1679 his competence was rewarded when he was given command of the frigate *Lark*. Virtually nothing else is known about Captain Collins's personal circumstances, although it does seem that while he was in favour at Court he did not have much of a financial income to sustain himself. The fact that he had never before had to deal with substantial sums of money possibly explains why he optimistically estimated that the British coastal survey could be accomplished for only £600!

For the purpose of his survey Collins was lent a handy 70ft cutter, the royal yacht *Merlin* (then another called *Monmouth*). However, the only tools provided for him were primitive, comprising a measuring chain, lead line and compass, a 5ft-radius quadrant for use ashore and a quadrant for hand-held use offshore. In 1681, armed with just those devices, he set out from Dover to work his way systematically along the south coast, laboriously putting landing parties ashore, taking bearings and making soundings. By late 1682 he had surveyed as far as Lands End, then in 1683 he covered the Scilly Isles and then mapped around the Severn estuary to Milford Haven and Dublin. He was already having financial troubles: progress was slower than he had envisaged, and all the while he was having to pay out for travelling charges, boat hire, horse hire, pilots, guards, assistants and other contingent expenses. Also he had naively ensnared himself in an unwise and financially confusing arrangement whereby he was partly funded by Trinity House; he received nothing from the Admiralty, and had no prospect

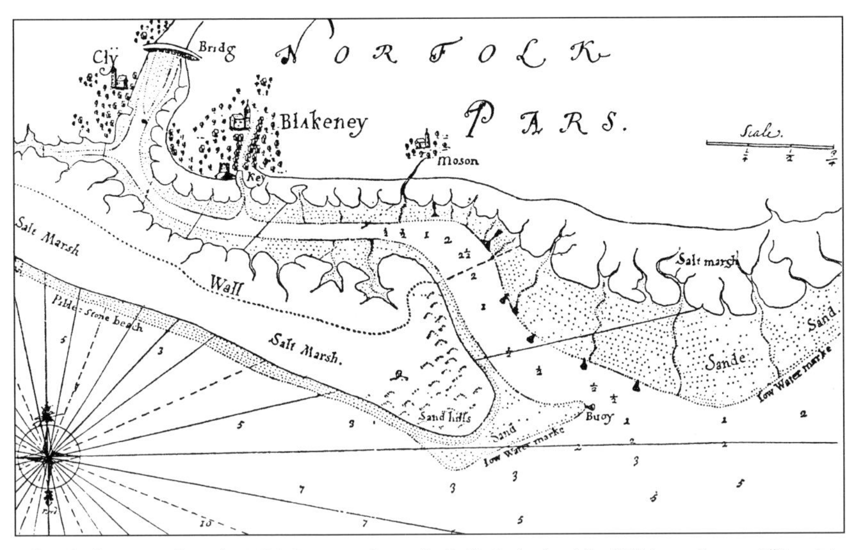

A hundred years on from the 1681–7 survey, Greenvile Collins's charts of the British coast were still in print, and much prized by mariners for their robust simplicity. (Robinson, *Marine Cartography in Britain*, 1962)

of being reimbursed by the Crown until the survey was complete. So he must have been feeling quite agitated in February 1683, when, according to the diarist John Evelyn, he arrived at the Court of St James to show the results of his previous two summers' work.

By then Collins had cleverly devised a rolling production system whereby he would somehow get his charts published as he did them, and it seems he occasionally managed to persuade local dignitaries to sponsor him as he moved along the coast – rather like a radio roadshow! Thus in 1684 when he surveyed from Harwich to Edinburgh and in 1685 when he moved on from Edinburgh to the Orkneys and Shetlands, such personages who got their names immortalised in print included the Duke of Norfolk, the Earl of Perth and even the Magistrates of the City of Aberdeen. By the end of 1687 he had finished the coasts around the rim of the Irish Sea, and the survey was complete at last.

Still unpaid by the Crown, and by now under considerable financial strain, Collins eventually found a bookseller called Richard Mount on Tower Hill, who agreed to publish his work. Collins had originally produced 120 harbour and coastal charts, but to reduce costs selected only 48 to be engraved; and in 1693 – twelve years after the enterprise had begun – *Great Britain's Coasting Pilot* went on sale to the general public in three leather-bound folio volumes.

Captain Collins must have wondered why he had bothered. Immediately the folios appeared influential critics condemned him for his primitive methods. Members of the Royal Society, in particular Robert Hooke, criticised him for being unscientific. Trinity House complained of the 'Ill performance of the Book of Charts', claiming Collins had relied too much on compass intersections. Even Pepys, who had been influential in his appointment, wanted to know why the Dutch were not copying them! Unhappily for Collins, the publication could scarcely have been worse timed. After the establishment of the Paris Observatory in 1667, a survey by triangulation of the coasts of France had been put in hand and the resultant charts appeared in *Le Neptune François* in 1693. Its beautifully produced and mathematically precise charts made Collins's coastal outlines seem like crumpled bags of potatoes in comparison.

Collins duly submitted an invoice to the tune of £1,914, which included the £1,400 total expenses incurred during the seven years he was out in all weathers and his wages. Trinity House reluctantly made him a grant of £50 and the Admiralty grudgingly offered him £1,414, so that despite all he had achieved Collins was £450 worse off than he had been in 1681. Although Charles II continued to have faith in Collins, and in January 1694 appointed him captain of his favourite royal yacht *Fubbs*, the financial shenanigans had so broken Collins's spirit that he died in March. Collins had been the first ever Royal Hydrographer (Alexander Dalrymple was the first

The Parallel Ruler

The precursor of the ubiquitous Breton plotter was the 'rolling' parallel ruler, invented by a master mariner called William Field. A simple instrument to make from bronze, brass or boxwood, it sold throughout the world. Though he had retired from the sea to become a farmer near Milford Haven, in 1833 Captain Field heard of a brig foundering on rocks off the Pembrokeshire coast – and swam out to the wreck to rescue the survivors. For his brave action the Lifeboat Service arranged for him to be awarded a silver medal.

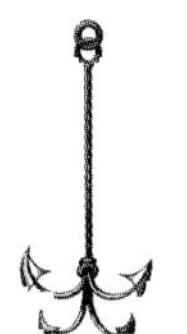

The modern Breton plotter evolved from Captain William Field's rolling parallel ruler. (Admiralty, *Manual of Seamanship*, 1932)

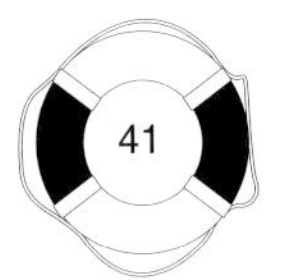

Hydrographer to the Admiralty (1795–1808)). His thorough survey was the first ever done of the British coast, and regardless of all the criticism his charts were so popular with British mariners for their robust reliability that they remained in print for the next one hundred years.

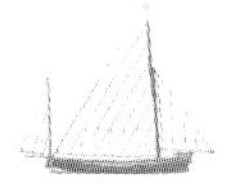

Hydrographic Office, www.ukho.gov.uk; reprints of old Admiralty charts are available from www.admiraltyleisure.co.uk.

John Campbell's Lunar Sextant

The quality of navigational instruments was revolutionised between the beginning and the end of the eighteenth century, and a particularly significant improvement was the new-fangled sextant. The first real ancestor of the modern-day sextant as a multi-purpose nautical instrument was the cross-staff, first described in 1342 by a scholar named Levi ben Gerson. An adaptation from an earlier astronomical survey device, it consisted of a frame (staff) over 30 inches long with scales engraved on all sides. Perpendicular to the frame ran two or more transoms or 'crosses' (hence the name). By lining up the horizon with one end of a cross and the celestial object with the other end, the observer had a simple trigonometric computer! The cross-staff was extremely useful in the art of navigation, since it embodied all the functions of recording the altitudes of the sun, stars, moon and planets, as well as terrestrial objects. Most frequently the observer's latitude was found by 'shooting' the sun, an expression popularised because of the resemblance of the instrument to the crossbows of the period. Even now celestial observations are still referred to as 'shots'.

A slightly more advanced cross-staff, invented by an English navigator called John Davies in the 1590s remained the principal navigational instrument until a reflecting quadrant was introduced by Hadley in 1731. In a few respects this was similar in appearance to the modern sextant. However, like Davis's cross-staff, Hadley's quadrant was used primarily for latitude observations. During Hadley's lifetime people came to realise that an alternative means of position finding could be achieved by accurately measuring the angle of the moon to a horizon. However, at that time charts showing the changing position of the moon were not accurate enough for longitude determination. The first reliable lunar tables

Campbell's 1757 lunar sextant enabled seafarers to determine their position accurately. (Rosser, *A Self Instructor in Navigation*, 1863)

were those developed by the German astronomer Tobias Mayer and published in 1753, and Mayer suggested the Hadley octant be enlarged to a full circle. Thus in 1756 the Board of Longitude commissioned the leading English navigational instrument maker, John Bird, to make a reflecting circle. That full-circle Hadley's quadrant was given to a Captain John Campbell RN to test.

Campbell was no inventor, but during the trials he conducted off the island of Ushant, near the Brittany coast, he proved himself to be a brilliant innovator. The reflecting circle had potential because it could measure angles greater than 90 degrees, but it was so heavy it had to be rested on a staff. He wondered if the device could not be made far less burdensome, but without forfeiting any efficiency, if it was cut down to measure 120 degrees. He arranged for Bird to make a reflecting instrument with a reduced arc, which he then satisfactorily tested in 1758–9. Initially this so called sextant was also quite heavy, being made of brass with a 20-inch radius overall. But once the basic principles had been perfected, lighter models became available.

Much of the subsequent commercial success of the sextant can be attributed to the Cambridge astronomer Nevil Maskelyne, who was eager to establish that the astronomical solution to the problem of longitude was superior to the mechanical – in other words, that lunar sextants mattered more than chronometers. In 1765, having recently been appointed Astronomer Royal, Maskelyne persuaded the Board of Longitude to sponsor a British *Nautical Almanac*. The first edition appeared in 1767, and subsequently the demand for sextants mushroomed. Production was much accelerated by the development of a dividing engine for machining scales, invented in London by Jesse Ramsden in about 1771. Ramsden's dividing engine was truly revolutionary insofar as it enabled sophisticated navigational instruments to be mass-produced. Makers no longer needed to make the scales, the most delicate part of the new

THE

NAUTICAL ALMANAC

AND

ASTRONOMICAL EPHEMERIS,

FOR THE YEAR 1768.

Publiſhed by ORDER of the

COMMISSIONERS OF LONGITUDE.

LONDON:

Printed by W. RICHARDSON and S. CLARK,
PRINTERS;
AND SOLD BY
J. NOURSE, in the Strand, and Meſſ. MOUNT and PAGE,
on Tower-Hill,
Bookſellers to the ſaid COMMISSIONERS.
M DCC LXVII.
[Price Two Shillings and Six Pence.]

By 1768 sights from Campbell's sextant could be calculated with *The Nautical Almanac*. (Maskelyne, *The Nautical Almanac*, 1768)

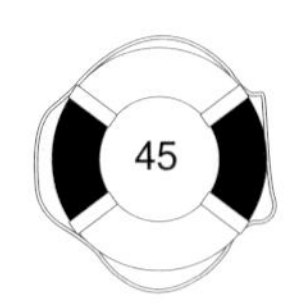

sextants, but could purchase them from Ramsden for just 3 shillings per instrument. Thus the sextant John Campbell had innovated became widely available and very popular.

Campbell's original sextant is exhibited at the National Maritime Museum, Greenwich, www.nmm.ac.uk. See also the Mystic Seaport Museum in the United States, www.mysticseaport.org.

Edward Massey's Distance Log

In 1802 a mechanical device was invented that would revolutionise the measurement of speed and distance travelled at sea. The earliest known attempt to estimate mileage at sea was made by the Roman engineer Vitruvius, who used a type of waterwheel fixed to the hull of a ship. A drum on board was filled with pebbles, and every time the wheel revolved one pebble fell out into a collection box. By counting the number of pebbles in the box an estimate of distance travelled could be obtained! It was quite an ingenious system, but sadly was soon entirely forgotten.

From the late sixteenth century two other methods of measuring distance run came into use. Dutch mariners favoured the Dutchman's Log. A piece of wood was dropped into the sea level

Nautical Superstitions

Good luck	Bad luck
Throw broomhead overboard	Lose mop overboard
Whistle when wind not blowing	Whistle when wind is blowing
Feather of a wren	Shoot an albatross
Pour wine on deck	Make a wineglass ring
Launch with champagne	Have plants on board

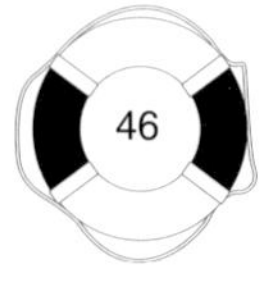

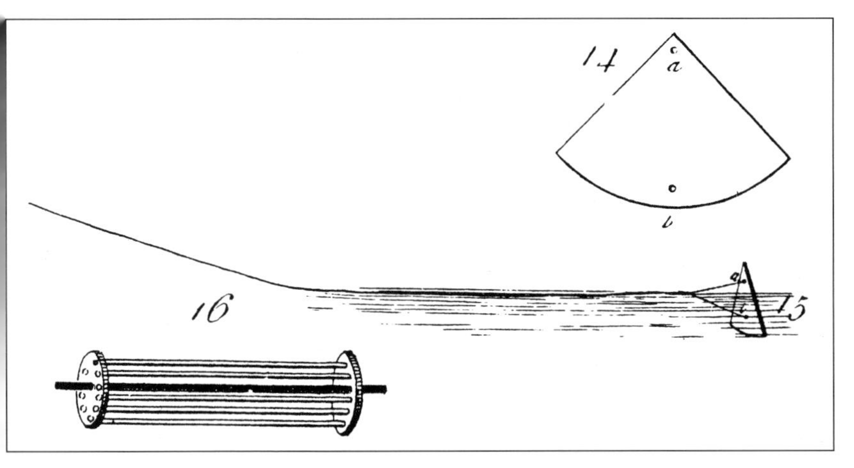

Traditionally a vessel's speed was timed with a chip log, by reeling out a wooden quadrant attached to a line marked with knots. (Falconer, *Marine Dictionary*, 1780)

with the bow, and timed as it floated between two predetermined marks on the side of the vessel. Knowing the distance between those points, it was an easy matter to calculate the rate of progress. Seasoned mariners could improvise by monitoring a passing bubble in the waves. However, the success of all such methods depended on the accuracy of observation and timing.

The 'Common' or 'Chip' Log, initially reported in 1574, usually consisted of a triangular wooden board called the log-chip, weighted at its base to keep it upright in the water, with two or three holes; the log-line was attached to this board by a crow's foot of cords, to one of which a release peg was attached. To use it, the chip log was thrown overboard, well clear and to windward of the ship, and the line was paid out from the reel on which it was wound. The first 10 or 20 fathoms of the log-line, marked by a piece of bunting, was called stray-line, and this served to leave the log clear of the disturbed wake of the vessel before timing was begun. The line was run out for 30 seconds, then measured in fathoms by the stretch of

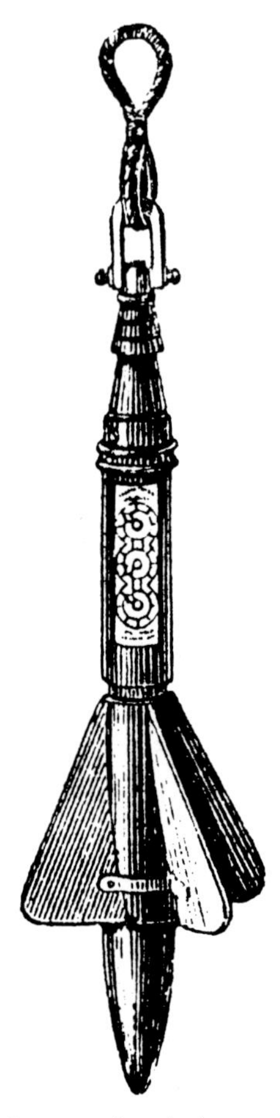

Accurate though the 15in Harpoon Log was, it required hauling in for its dials to be read. (Neison, Kemp and Davies, *Practical Boat Building and Sailing*, 1900)

arms. According to Kevin Desmond's *Guinness Book of Motorboating*, by calculating these results, the speed of the ship in leagues per hour could thereby be obtained.

With the growth of seaborne trade in the seventeenth century various inventors responded to the need to find a more accurate measurement of a ship's speed. Some form of mechanical log seemed to be required. Among the earliest was one designed by Robert Hooke, which was demonstrated to the Royal Society in 1683, although there is no record of whether it was tested in a ship. In 1754 the nautical engineer John Smeaton attempted to produce a lightweight rotator. The most promising was that suggested by William Foxon of London, who in 1772 patented a mechanical log. This consisted of a helical rotator towed behind a ship, and linked by the log-line to an inboard movement. One dial recorded the distance run, and another was used in conjunction with a half-minute glass to measure the ship's speed. However, as was the case with all the others, this system suffered from excessive friction, which distorted the figures.

The first commercially successful cumulative recording log was patented by Edward Massey in 1802. Called the Perpetual Log because it was constantly in motion (unlike the intermittently functioning Common Log), it consisted of a shallow rectangular box on a dart-shaped float plate. The box contained the registering wheelwork, which was connected by a universal joint to a pointed metal tube with flat metal vanes set at an angle. The entire mechanism was towed astern of the ship, and had to be hauled in, read and reset at each alteration of course. Nevertheless the results proved impressively accurate and the Massey Log was extensively used at sea for much of the nineteenth century.

Of Massey himself not much is known, so there is no means of telling how he developed his design. But it was his nephew Thomas Walker who took the next significant step in the evolution of the mechanical log when he introduced the Harpoon Log in 1861. This was an improvement on the Massey Log in that the rotator and registering mechanism were contained in a single unit, although the log still had to be hauled in when readings were required.

In 1876 Thomas's son Thomas Ferdinand Walker patented his Cherub taffrail log, which consisted of a towed rotator connected by a log-line to a register mounted on the taffrail. This could also be used as a speedometer! The advantage of the Walker taffrail log was that it did not need to be hauled in for every reading, although like all towed mechanical logs it had certain limitations. It was vulnerable to any accident – even a slight alteration in the angles of the vanes would have a significant effect on the accuracy of the results – and in rough weather it could give false readings. Nevertheless, the Walker Log, a

For many decades the trailing Walker 'Excelsior' was a yachting favourite. (*Yachting Monthly*, March 1955)

descendant of Edward Massey's original invention, became a classic piece of equipment which remained in production for almost 125 years.

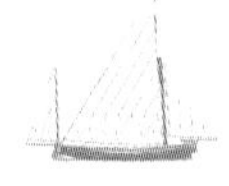

There are various speed and distance logs on show at the National Maritime Museum, Greenwich, www.nmm.ac.uk.

Francis Crow and the Ship's Compass

The earliest known form of compass appeared in 2634 BC during an uprising led by a dissident Chinese prince, known as Chiyou, against the Emperor Huang-ti. When their armies eventually met the emperor ordered his troops to attack, but a mist rolled down causing such confusion that it several times enabled the enemy formations to escape. Eventually the emperor summoned forth a 'south-pointing' chariot to guide his forces through the mist. His army charged behind this vehicle and duly defeated the rebels. It seems the hand of the figure on the chariot always pointed south, but how it did so remains a puzzle. It probably used a system of differential gears connecting the magnetic figure to the chariot wheels.

Centuries later that device was superseded by simple compasses which consisted of a lodestone and a dry card marked with cardinal points, although the major disadvantage of such a type of compass was the tendency of the card to jam or swing about erratically in bad weather. In such conditions mariners had to rely on that precarious tradition of navigation known as 'by guess and by God'.

What reputedly alerted Francis Crow to the possibility of improvement was a remarkable story he happened to hear about a German coasting vessel in 1779 whose compass card was oscillating wildly until a sea broke on board filling the compass bowl – which then miraculously became steady. A silversmith with a workshop in Faversham, Kent, Crow evidently had some experience of compass-

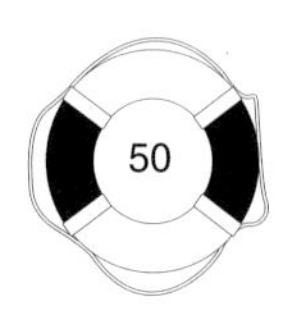

The earliest form of Chinese compass was the 'south-pointing' chariot. (Dixon, *'Yachting World' Annual*, 1967)

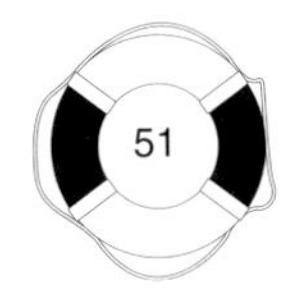

making, having patented a marine compass in 1801. Inspired by the captain's story, he set to work and in 1812 produced a revolutionary form of liquid compass. Its bowl contained spirit – so it would not freeze – to damp the movement of the card, and a float which rested against an inverted pivot below the glass was attached to the card to reduce weight and friction. Crow also designed an air-chamber in the top of the bowl which later he fitted with a form of leather valve to allow for expansion of the liquid.

Crow may not have been the first instrument maker to create a liquid compass in a metal bowl – in 1781 Captain Sir William Challoner had tried just such a device made by a Londoner called Barton. But it does seem that Crow's was the first to have a discernibly modern appearance. In 1813, the year it was patented, Crow fitted to the bowl of his liquid compass an azimuth prism – innovated by Charles Schmalcalder only a few months earlier – and thereby enabled it to function as a bearing compass as well as a basic steering compass.

Yet Crow lost out. Captain John Ross took six compasses on his voyage of Arctic exploration in 1818, including Crow's (seemingly the original 1813 model), and reported favourably on it. He 'found it to be very good save that the steering compass card was too heavy'. Surprisingly, the Admiralty took no further interest in Crow's compass, which seems to have been misfiled as a specialist instrument of discovery. Disillusioned, Crow busied himself with other inventions, most notably a maritime lightning conductor. In 1820 he moved to Greenwich and later to Chelsea, where he died in 1835. The task of refining the liquid compass fell to others, notably Grant Preston, who in about 1832 attempted to solve the problem of bubbles forming in the liquid by fitting reservoirs above the bowl in which the liquid could rise and fall with changes of temperature. But by 1837 the state of the Royal Navy's card compasses was so dire that a committee was appointed to study their condition. One of the results of their deliberations was the design which appeared in

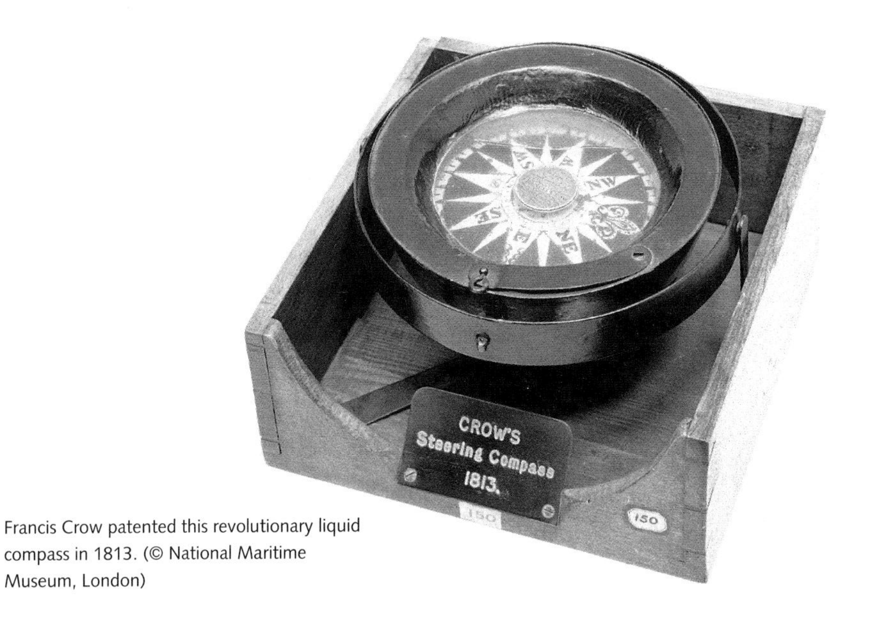

Francis Crow patented this revolutionary liquid compass in 1813. (© National Maritime Museum, London)

1842 of a new standard card compass which consisted of four or five parallel straight strips of magnetised steel fixed under a card. Sadly, the review missed the opportunity officially to develop liquid compasses for ships. Even though in 1845 the Admiralty issued an order that every Royal Navy ship was to be equipped with a liquid compass, it was only for steering by in rough weather.

Eventually the liquid ship's compass was developed in Boston, Massachusetts, by Edward Ritchie, a former ship's chandler who was winning renown as a physicist. In 1861, while working as an instrument maker on an order for the US Navy for Admiralty-style liquid compasses (which he frequently repaired), Ritchie learnt of a decision by the Superintendent of the Naval Observatory to encourage American compass innovations. Ritchie's initial interpretation of that was limited to finding a means of preventing compass needles from oxidising. Then in July 1862 he had a radical thought. Why not do away with the problem altogether by encasing the needle in a floating

Reckoned to be more reliable in heavy seas than traditional compasses, liquid compasses became standard equipment on lifeboats. (Lewis, *The Lifeboat and its Work*, 1874)

ring? The Superintendent, Captain James Gillis, endorsed the idea, and Ritchie perfected it further using a cross float within the tubular ring. Widely adopted by the US Navy, and later by the American merchant fleet, this compass was one of the most advanced designs to date. By 1866 Edward Ritchie's compass was being made under licence in France. Yet when he travelled to England to sell the idea to the Admiralty, he met with no success.

By then, there had been some progress in Britain towards developing a liquid compass for boats. In 1833 liquid compasses began to be produced by a firm of chronometer-makers in the Strand called Edward Dent. A newer patented version was then introduced which was claimed not to suffer from the problems of discoloration inside the bowl that had plagued the early liquid compasses. Dent's had traced that fault to the unwise use of cork gaskets, replacing them with rubber ones. Dent's fluid compass was particularly highly thought of by the Royal National Lifeboat Institution, for whom the company additionally supplied a japanned copper binnacle containing double gimbals and even an oil lamp reflector. Yachtsmen were soon able to buy a similarly compact compass and binnacle from Henry Hughes & Sons, of Fenchurch Street, while for use in emergencies, or by cost-conscious amateur sailors, the famously robust British Army liquid-filled Francis Barker marching compass was introduced in 1848.

However, the Admiralty remained determined to put their faith in the card compass and duly accepted an enormously sophisticated type invented by William Thompson. This was housed in a special binnacle containing magnets and spheres for the correction of errors caused by the steel from which ships were built. During the bombardment of Alexandria in 1882 the system had remained accurate even when the British warships were subject to acute vibration. In 1901 a greatly improved patent compass was made by John Dobbie of Glasgow which was of such steadiness that in 1906 it enabled the Royal Navy to switch over fully to liquid compasses using a Dobbie design perfected by the head of the Admiralty's Compass Branch, Commander L.W.P. Chetwynd. However, it was not until the Second World War that a device called the Sestrel hand-bearing compass was developed for the Navy's smaller fighting vessels. Subsequently used by yachtsmen, this device was remarkably similar to the azimuth compass Francis Crow had patented back in 1813!

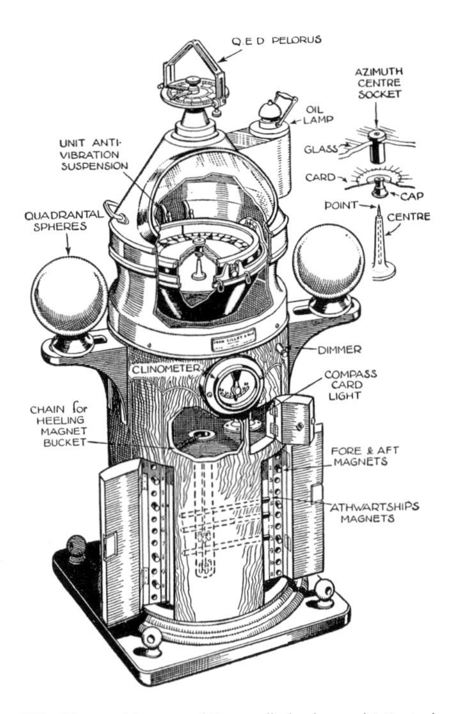

Warships and large yachts usually had a sophisticated Kelvin card compass, but in the 1900s that too was widely replaced by the liquid compass. (Carr, *The Yacht Master's Guide*, 1940)

An early Francis Crow compass is at the National Maritime Museum, Greenwich, www.nmm.ac.uk; the Ritchie Corporation's website is www.ritchienavigation.com.

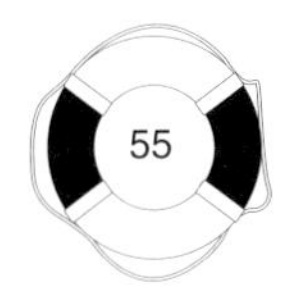

Stove Gimbals

Gimbals had been in existence for centuries, but no one had thought of attaching them to a yacht stove until the late nineteenth century, when Frank Rippingille of Birmingham did just that with his patent gimballed Rippingille stove. In 1896 chandlers were selling these swinging stoves for 30 shillings.

William Thomson's Depth Sounder

Experiments made in June 1872 on board a British sailing yacht called *Lalla Rookh* began the process of breaking down the centuries-old method of depth sounding. Traditionally depth sounding on ships had been done from a platform called the 'chains' on either side of the forecastle; where soundings needed to be taken, a leadsman, or a team of leadsmen, would be stationed here to 'heave the lead'. If no platform was fitted the process could be carried out from the sea-boats, or from any position along the vessel's side. The weight itself was usually available in three sizes: a 5–7lb boat lead for coastal waters, a 14lb hand lead for deeper waters, and a 28lb lead for deep sea use. The underside of the lead was hollowed out and might be filled with grease, the purpose of which was to obtain a specimen of the sea floor. With a good swing the blue lead could be made to emit a cooing sound rather like a pigeon, hence the expression 'blue pigeon'.

Sounding was a specialist skill and required careful procedural techniques. The line had to be hauled in with the right hand under and the left hand over, so as to make a right-handed coil and prevent it snarling. Also the lead always needed to be hove in the direction in which the ship was moving. For specialist advice yachtsmen could consult the Admiralty's *Manual of Seamanship*, which advised: 'When calling soundings: if the depth of water obtained corresponds with any of the marks on the line a leadsman would call "By the mark", "By the mark 7", &c. If a quarter of half a fathom more than a mark, he would call "and a quarter 7", "and a quarter 8", &c., or less than a mark or a deep, then he calls "a quarter less 7", or "a quarter less 8", &c.' The line itself was made of hemp and for boats was usually 10 to 12 fathoms long, marked in feet up to 4 fathoms. For larger vessels the line could be long as 300 feet, and could have various forms of markers, some of which varied in shape so they could be distinguished at night. The standard symbols were:

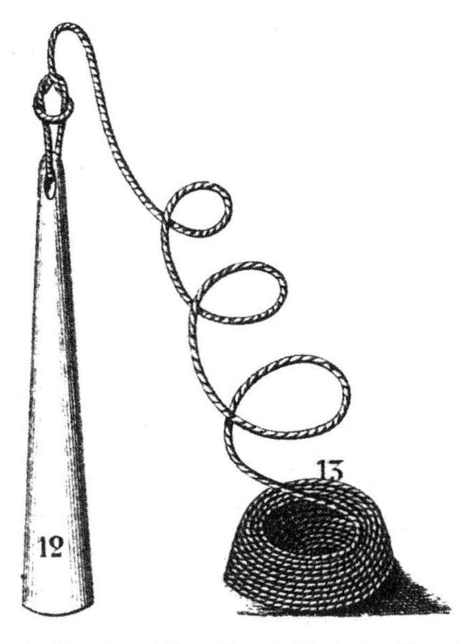

Traditional lead line. (Paasch, *Illustrated Marine Encyclopedia*, 1890)

Marks	*Fathoms*
Leather with two ends	2
Leather with three ends	3
White cotton	5
Red bunting	7
Leather with hole through	10
Blue serge	13
White cotton	15
Red bunting	17
Strand with two knots	20

To complicate this archaic system further a line longer than 25 fathoms was marked with a knot every 5 fathoms, and the intermediate fathoms ('deeps') on the standard line were unmarked.

Lalla Rookh was owned by Sir William Thomson, an eminent physicist in the field of transmission of electrical currents, who had worked on the transatlantic telegraph cable, where the process of depth sounding was an extremely laborious one. In 1870, after the death of his first wife, he spent much of his spare time on *Lalla Rookh* cruising to distant places, and he used the yacht as a boffins' floating laboratory for developing high-tech ingenious experimental devices such as a card compass. Thomson reckoned that the main problem with depth sounding was the thickness of the rope: for a very heavy sinker, a thick rope some 1½ inches in circumference was needed, although it clearly offered considerable resistance to motion through the water and took a long time to reach the bottom of the sea. For this reason the ship might have to heave to while the line ran out and while it was being heaved back in. Hence sounding, other than the use of the hand lead in relatively shallow water, was not routinely used as a

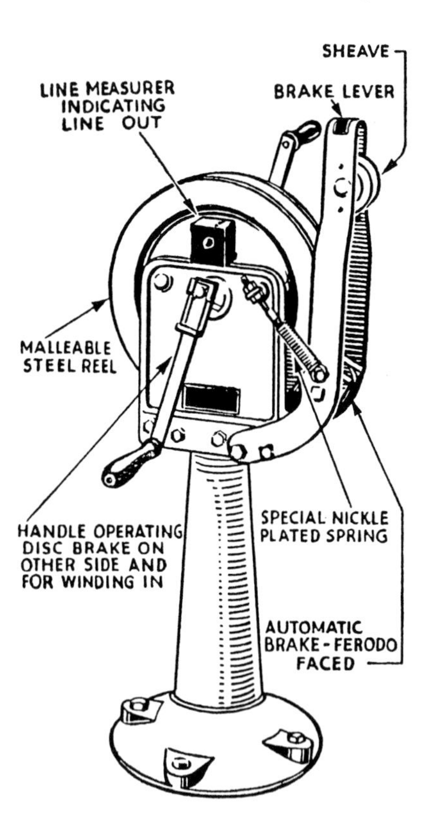

The 'Sceptre' sounding machine similar to that devised in 1872 by William Thomson. (Carr, *The Yachtmaster's Guide*, 1940)

navigation device, notwithstanding the importance of the information it could give when a vessel was approaching land.

To enable the sounding line to slip down more quickly Thomson chose to use piano wire. In 1872 he demonstrated the practicability of this in the Bay of Biscay where he sounded to a depth of 2,700 fathoms with a 30-pound lead hung from a line made up of sections of piano wire spliced together. He then invented a depth measurer, which consisted of a glass tube closed at the top and coated inside with chromate of silver, which is discolored by the action of sea-water. As the contraption descended, the increased pressure forced the sea-water to compress the air, thereby precisely indicating the depth by the height of the discoloration. To control the process he invented an electric winding machine with a drum for reeling in and dunking the lead. With two such devices boomed out on spars on either side of a ship's bridge a vessel could sail at a speed of 10 knots in 20 fathoms of water and obtain soundings once every half-minute.

However, Thomson's *Lalla Rookh* was a substantial vessel of 126 tons! So unless yachtsmen wished to burden themselves with the hefty Thomson winding equipment they had to wait until a different and more compact form of the device came on the market. As early as 1822, in Lake Geneva, Switzerland, Daniel Colloden had experimented with an underwater bell in an attempt to calculate the speed of sound underwater. But in 1911 it was Dr Alexander Behn of Kiel who first reckoned it might be possible to calculate water depth by timing echoes. This theory was further developed by a French physician called Paul Langevin by means of dropping explosives. Before his death in 1907, William Thomson, who was made Lord Kelvin in 1892 for his services to marine navigation, had established a marine equipment business, which in 1941 merged with a firm of instrument makers to form Kelvin Hughes. In 1925 Henry Hughes & Sons had pioneered the use of an electronic echo-sounding system for the Royal Navy and by 1931 had produced a high-speed boat gear, suitable for hydrographic surveying, which was subsequently evolved by other companies into the Seagraph and later the transistorised Hecla. All this meant that by the late 1950s most yachtsmen had available to them an echo-sounder that was affordable, reliable and easier to use than Thomson's improved boat lead.

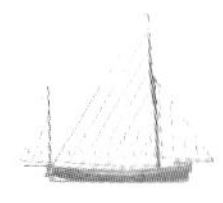

There are various examples of depth sounding devices at the National Maritime Museum, Greenwich, www.nmm.ac.uk; see also www.kelvinhughes.co.uk

Ernst Abbe's Binoculars

Until the invention of revolutionary fieldglasses in 1893 by the Carl Zeiss Company, the only binoculars available to yachtsmen were variants of simple opera glasses. Binoculars themselves had first been invented in Holland by Hans Lippershey as long ago as 1608, and for a while in the 1850s there had been some likelihood that a quantum leap forward in optical engineering would be made by an Italian artillery officer called Ignaz Porro. By 1839, in his efforts to devise better range-finding equipment, Porro had invented a rapid surveying-type geodetic instrument called a tachymeter, and had gone on to suggest that tele-lenses could be used to photograph

The 'Grabbit' Boathook

First on sale in 1913, the 'Grabbit' patent boathook soon proved ideal for short-handed crews, and an enthusiastic endorsement in *The Corinthian Yachtsman's Handbook* by the cruising pioneer Francis Cooke encouraged its success.

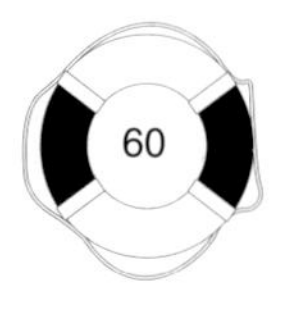

distant buildings. Porro's most promising breakthrough in optics came in 1850 when he reckoned that the use of prisms might not only improve the performance of telescopes, but also enable them to be more compact. He also believed he had the means to produce such 'Monoculars', since having left the army in 1842 he had been influential in founding the Institut Technomatique in Paris, and had also opened his own workshop in Turin. Nevertheless the monocular failed since he lacked the manufacturing procedures needed to create prisms of the necessary precision, and also because of the poor quality of the glass available.

Other optical makers also had a go at making monoculars, and a Parisian called A.A. Boulanger even went so far as to try to make a binocular version of Porro's prismatic telescope. All these attempts failed to produced prisms of the required quality and so were unsuccessful. However, apparently without any knowledge of Porro's work, a hitherto obscure academic called Ernst Abbe built an experimental prismatic telescope at the University of Jena in 1873, and then produced another version in 1879, which was also never marketed. Determined to find a means to improve the quality of his materials, he unsuccessfully attempted to make glass using fluorspar. In 1879 Abbe sought the assistance of an experimental chemist called Otto Schott, who was already experimenting with a fluorite-type glass. Several years after helping Schott to found his company, Abbe's perseverance was rewarded when the Schott smelters produced an exceedingly clear striae-free fluorite crown glass. This resulted in a monocular far superior to anything seen so far.

Fortunately, since 1875, Abbe had progressed from a technical adviser to a business partner with Carl Zeiss at Jena, who were manufacturers of optical instruments. Zeiss himself was keen to employ Abbe for his theoretical knowledge because, since founding his first workshop in Jena in 1846, he had been wanting to break away from the customary trial-and-error approach in the manufacture of precision optical instruments. By the early 1890s Zeiss's experience in the mass-production of quality microscopes, and later photographic lenses, made them ideal candidates for manufacturing the improved monocular. But in 1893 they encounted serious difficulties in preparing to register the contraption at the German Patent Office: the 'invention' of the image-reversing prism was not unique as Porro had thought of it first, so the patent application was turned down. To get round this problem Zeiss's ingeniously bundled the improved monoculars into pairs to form binoculars, but as binoculars were not unique either, they further devised a brilliant system of hinging together the telescopic cylinders, which hitherto had always been unalterably fixed. Thus in October 1894 a patent was granted for a 'double telescope increased objective separation' – effectively the world's first adjustable prismatic binoculars.

Called *Feldstecher* ('fieldglasses'), the new Zeiss prismatic binoculars went on sale in 1895, and were recognisably better quality than the usual Galilean 'opera' glasses. Abbe was convinced that the public would buy such handy devices despite their weight – they were heavy because of the weight of the prisms – and he was right. They were compact and easy to handle, while the large distance between the lenses, caused by the stepped arrangement of the prisms, allowed the generation of an image with an improved impression of depth. A range of models was produced, with various magnifications (of 4, 6 and 8) and customers were undeterred by the fact that they were two or three times more expensive than traditional opera glasses. In addition to yachting, they were also in demand for birdwatching and astronomy. Sportsmen outside Germany were keen to acquire them so Zeiss's were soon making money-spinning licensing agreements abroad to enable the new binoculars to be made in Austria, Russia and even as far afield as America. They proved so popular that rival German manufacturers jumped on the bandwagon, bypassing the patent restrictions by making binoculars without the distinctive Zeiss hinged mechanism.

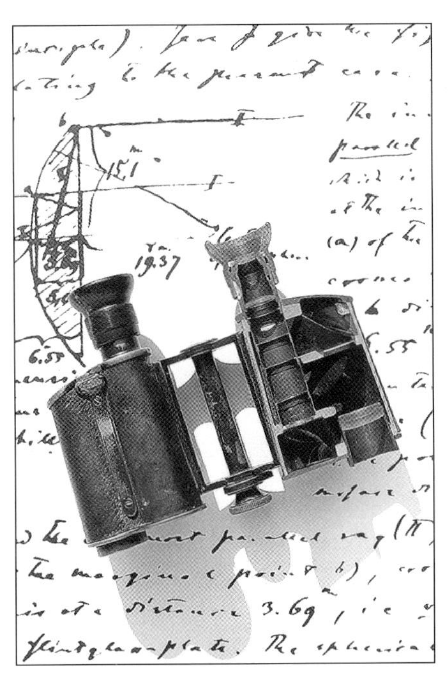

Ernst Abbe's calculations produced the revolutionary Zeiss prismatic binoculars in 1893. (Zeiss)

What alerted the British armed forces to the tactical threat posed by the Zeiss binoculars was their use in the 1899–1902 Boer War. One particularly effective model was the 6 x 21 binocular (meaning it magnified 6 times and had a 21mm objective lens). Weighing only 2lb, it could be kept in the pocket of a trench coat. Eventually increased to 6 x 24, this model could be bought at the Carl Zeiss shop in London, which had been established just off Regent Street in 1894. During the First World War both the Regent Street shop and the Zeiss manufacturing plant in Mill Hill were forcibly taken over by Ross Ltd of Clapham to make Zeiss-style 6 x 24 binoculars.

However, by 1915 the Jena concern was already a step ahead with their 7 x 50 *Doppelfrenrohr* binoculars which were issued to the German Navy in that year. They were formidable because of their excellent light efficiency performance. In the interwar years the Admiralty chose not to equip the Royal Navy with the Zeiss 7 x 50s, and in 1930 commissioned a Glasgow optical company, Barr & Stroud, to develop a British equivalent, known as the CF 30. Nevertheless, many naval personnel would still have preferred to use

The Dorade Ventilator

In 1933 the American designer Rod Stephens devised an ingenious ventilator with an air intake box encasing a tube which allowed air to enter while keeping water out. First installed on the sensational yawl *Dorade* which won the Fastnet in that year, the new ventilator became known as the 'Dorade box' when it was fitted on other yachts.

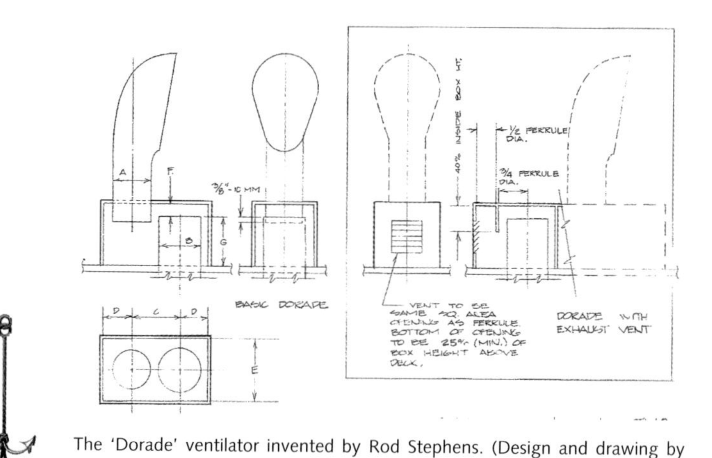

The 'Dorade' ventilator invented by Rod Stephens. (Design and drawing by Sparkman and Stephens copyright © 2004 by Sparkman and Stephens, Inc.)

Zeiss products. In 1935 the performance of the *Doppelfrenrohr* was further improved when the Zeiss laboratories secretly invented a revolutionary anti-reflection coating which increased their light transmission and thereby offered better observation in poor light conditions. During the 1938 Munich Crisis the Commander-in-Chief Home Fleet, Admiral Sir Charles Forbes, was so concerned by the technology gap that he believed a hypothetical German lookout using Zeiss 7 x 50 binoculars would spot an enemy vessel more quickly than a British opponent using Barr & Stroud 7 x 50s; he even wondered if a sea battle's outcome might be determined by the

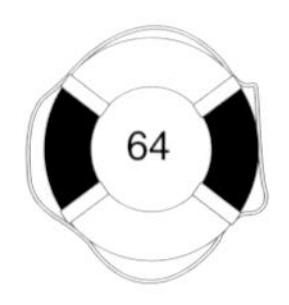

standard of optical instruments used by the rival fleets. The real quality of the improved Zeiss 7 x 50s remained unknown until August 1941 when the submarine *U-570* surrendered off Iceland and her pressure-proof binoculars were sent for evaluation tests – which established they were optically superior.

In the aftermath of the two world wars thousands of high-quality surplus Zeiss binoculars became available for yachtsmen to purchase. Ernst Abbe would have been horrified by such conflicts. When Zeiss himself died in 1888, Abbe became the guiding light in the development of the Carl Zeiss Foundation, a charity whose very purpose was to facilitate human progress by advancing the German optical industry.

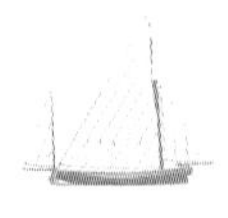

Carl Zeiss International, www.zeiss.com; and the Optical Museum, Jena, www.ernst-abbe-stiftung.de.

3

Weather and Signals

Luke Howard's Cloud Classification

The inspiration for the devising of a revolutionary classification of clouds was the presence in the skies of the northern hemisphere in 1783 of the 'Great Fogg', a haze composed of dust and ash from a number of violent volcanic eruptions in Iceland and Japan. At the time Luke Howard was an eleven-year-old living in London, and he became fascinated by clouds after seeing the stunning sunsets created by the accumulation of dust in the air. On the evening of 18 August that summer his witnessing of the flight across the European sky of a stupendously fiery meteor crystallised his sense of vocation.

Luke Howard was the son of the inventive tinsmith Robert Howard, who introduced the newest technology Argand oil lamps to Britain. A hard-working Quaker pharmacist, by the time he was twenty-three Luke had a shop in Fleet Street, and later would pioneer the supply of quinine, newly isolated in France. Fascinated by wildlife, in 1800 he read a paper to the Linnaean Society on the subject of the classification of types of pollen. However, his real enthusiasm was for studying clouds – so much so that he began keeping a daily record of all that he saw in the sky on his journeys between Fleet Street and Plaistow in Essex where he had a factory manufacturing chemicals.

Luke Howard (1772–1864), the cloud classifier. (Science Museum/Science and Society Picture Library)

These observations he presented as a paper in 1803. He had joined a group of Quaker amateur scientists, known than as 'natural philosophers', who called themselves the Askesians (searchers after knowledge). Members took turns to read a scientific paper to the others. Luke Howard's turn came early that winter. His paper, 'On the modification of clouds', stunned the audience and was so well received it was published, appearing in the *Philosophical Magazine* in 1803 as 'Notes on the Modifications of the Clouds', illustrated with Howard's own watercolours. It eventually became a classic.

Essentially Howard had applied Linnaeus's method of classification to the varying forms of clouds. He had noted that there were three basic shapes: heaps of separated cloud masses with flat bottoms and cauliflower tops, which he named cumulus (Latin for 'heap'); layers of cloud much wider than they were thick, like a blanket or a mattress, which he named stratus (Latin for 'layer'); and wispy curls, like hair, which he called cirrus (Latin for 'curl'). To clouds generating precipitation, he gave the name nimbus (Latin for 'rain'). In addition to these four chief types, occurring in three layers in the lower atmosphere, he identified three intermediate discernible modifications. Thus there were, he claimed, twelve major cloud types:

Heaps: Cumulus family
- Fair weather cumulus
- Swelling cumulus
- Cumulus congestus

Layers: Stratus family
- Stratus
- Cirrostratus
- Cirrus

Layered heaps:
- Stratocumulus
- Altocumulus
- Cirrocumulus

Precipitating clouds:
- Cumulonimbus
- Altostratus
- Nimbostratus

These terms came to be generally adopted by meteorologists. But Howard did not stop there. Continuing with his routine observations of the capital's weather, in 1820 he published the *Climate of London*. Far ahead of his time, Howard was virtually the first person to realise that cities have an effect on the climate. By noticing that cities were often warmer than the countryside, he had in effect identified what came to be known as the 'urban heat island effect'.

For his accomplishments as an amateur meteorologist in 1821 Howard was made a Fellow of the Royal Society. But from there on, until his death in Tottenham in 1864, he received scant other public recognition. Not that he minded, since he was much concerned with his philanthropic and religious works. In that capacity, for example, he wrote tracts against profane swearing and was a zealous worker in the anti-slavery cause. He had earlier sponsored a movement for the relief of German peasants in the districts ravaged by the wars after Napoleon's retreat from Moscow.

Howard had been lucky as well as skilled in his cloud classification scheme. The famous physicist Robert Hooke had once considered devising a nomenclature for them, as had Aristotle many centuries earlier, but had never pursued the idea. Thus Howard received the honour because of his perseverance and dedication.

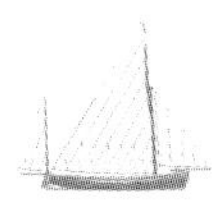

There is an English Heritage blue plaque outside Luke Howard's former home at 7 Bruce Grove, Tottenham, London N17.

Francis Beaufort and the Wind Scale

The idea of devising some form of code to describe sea conditions formulated in Francis Beaufort's mind in June 1805, when he was appointed captain of the armed naval storeship *Woolwich*. He wanted to make the ship's log more concise and more comprehensive.

NAVAL FOOD AND DRINK

FOOD	
Pusser's ham	Tinned corn beef
Cheesy-hammy-eggy-topside	Toast with cheese, ham and egg
Critter fritter	Fried mystery meat
Potmess	Stew
Sea-pie	Fish and vegetables covered with mashed potato
Spithead pheasant	A kipper
DRINK	
Blonde & Bitter	Coffee with cream
Flip	Pusser's rum and beer
Ki	Cocoa
Limey	Pusser's rum and lime
P&P	Pusser's rum and Pepsi
Skillygolee	Oatmeal and sugar

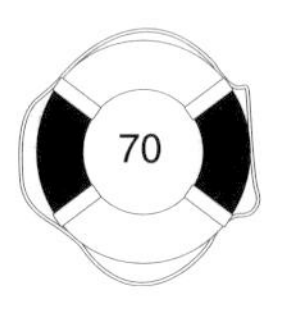

Admiral Francis Beaufort invented the Beaufort wind scale in 1806. (Met Office)

An Irishman, born in Navan, County Meath, Beaufort's father was a clergyman and a topographer of some distinction, who had published one of the earliest detailed maps of Ireland. Francis studied astronomy in Dublin before going to sea with the East India Company, later joining the Royal Navy. As a consequence of being shipwrecked because of poor charts, Beaufort became obsessed with the importance of education and the development of accurate charts for those risking their lives at sea. His naval career seemed to be over in 1800 when he received multiple wounds during a courageous attack off Malaga, and he spent the next few years recuperating – and establishing a telegraph line across Ireland.

From his earliest years at sea Beaufort had always kept a journal in which he made a note of the wind every 24 hours. Later on he recorded the weather every 2 hours. According to a biography of Beaufort by Nicholas Courtney, it seems that the basis for what eventually became known as the Beaufort scale was a table of wind speeds devised and published in 1779 in a *Memoir* by Alexander Dalrymple, who became Hydrographer of the Navy in Beaufort's time. What Beaufort did in 1806 was to increase Dalrymple's scale to thirteen wind states (ranging from 0 Calm to 13 Storm) by adding a new category, 'moderate breeze'. However, what made Beaufort's wind scale so unique was that it was based not on *measured* windspeed, which was difficult to ascertain because the necessary recording instruments were so unreliable, but on *estimated* wind force. This was something anyone could reckon simply by looking at the state of the sea and the surrounding conditions. Beaufort continued to use his classifications when writing in his log, but the system was first used officially by Robert FitzRoy in 1831 and was adopted by the Admiralty in 1838.

By then Beaufort had shown other accomplishments. He was badly wounded again in 1812 while surveying the coast of Asia Minor, but the maps he produced of the hitherto uncharted coast there were acknowledged as the finest yet submitted to the Admiralty. Returning home, he became a member of both the Royal Society and the Geological Society, and was a founder member of the Royal Geographical Society. Appointed Hydrographer to the Navy in 1829, he held the post for an astonishing twenty-six years. During that time he transformed it from a fairly dusty and obscure office to a busy hub responsible for directing some of the major maritime explorations and experiments of that period. For eight years Beaufort headed the Arctic Council during its search for the lost Arctic explorer Sir John Franklin. Overcoming many objections, Beaufort obtained government support for James Clark Ross's Antarctic voyage of 1839–43, intended to carry out extensive measurement of terrestrial magnetism. In addition he established an Admiralty Compass Department, produced a *Manual of Scientific Enquiry* to guide medical officers afloat, presided over a 'Grand Survey of the British Isles', and saw to it that a system of issuing *Notices to Mariners* was begun. But all these achievements were overshadowed by the lasting fame he won through the widespread use of his Beaufort wind scale.

The BBC Weather Centre, www.bbc.co.uk/weather, and the Meteorological Office, www.meto.gov.uk.

The Windspeed Anemometer

The hemispherical cup anemometer was invented in 1846 by Dr Thomas Robinson of the Armagh Observatory, and consisted of four hemispherical cups. The cups rotated with the wind, and a simple cogwheel mechanism recorded the number of revolutions in a given time.

Frederick Marryat's Signal Flags

The main advocate of a code of signals comprehensible to both warships and merchant ships was the author of the novel *The Children of the New Forest*. Born in 1792, Frederick Marryat had in his early years been a heroic naval officer. He had served on Admiral Thomas Cochrane's famous frigate *Imperieuse* and had frequently been involved in adventurous escapades such as attacking coastal vessels, and storming and destroying telegraph stations. Moreover he made it his speciality to dive into the sea to rescue anyone who fell overboard. His bravery for saving more than a dozen lives was rewarded with a gold medal from the Royal Humane Society. In 1819 he was made commander of the sloop *Beaver*, which was a guardship at St Helena until the death of Napoleon in 1821.

In 1803 Admiral Sir Home Popham had modified an existing form of naval flag signalling into a code called the *Marine Vocabulary*, which the Admiralty adopted for use in British warships, and indeed it was subsequently used for many years. However, Marryat realised that there was no such means of signalling for merchant ships, which could only communicate by hailing or by lowering a boat. He was not the first to see the need for a signal book: between 1814 and 1855 there was a veritable

Frederick Marryat, the originator of a basic signal flag code, was also the novelist who wrote *The Children of the New Forest*. (By courtesy of the National Portrait Gallery)

spate of such productions, mostly by former naval characters. But in 1817 Marryat produced his *Code of Signals for the Merchant Service*, which in due course acquired a tremendous reputation. Unlike the Popham system, which needed several numeral flags per hoist to denote a word, the Marryat system relied more on alphabetical flags. This was much more reliable, and proved so popular among master mariners that new issues of it continued to be published until 1879, when it reached its nineteenth edition. It was also used by yachtsmen, and indeed even served as the basis for the special *Universal Yacht Signals* code published by the Royal Yacht Squadron in 1847.

Marryat left the Royal Navy to take up a new career writing novels at Langham, his house and farm in Norfolk. His output was prodigious – some two or three books a year – the most popular concerned his fictional hero Peter Simple. His numerous children's books also did well. Nevertheless, having lost a fortune on a disastrous West Indies property investment, in 1847 he applied for service afloat with the Admiralty. Even then his signal flag code was still going strong, and it continued to be widely used by mariners several years after it was officially replaced with the International Code of Signals in 1857.

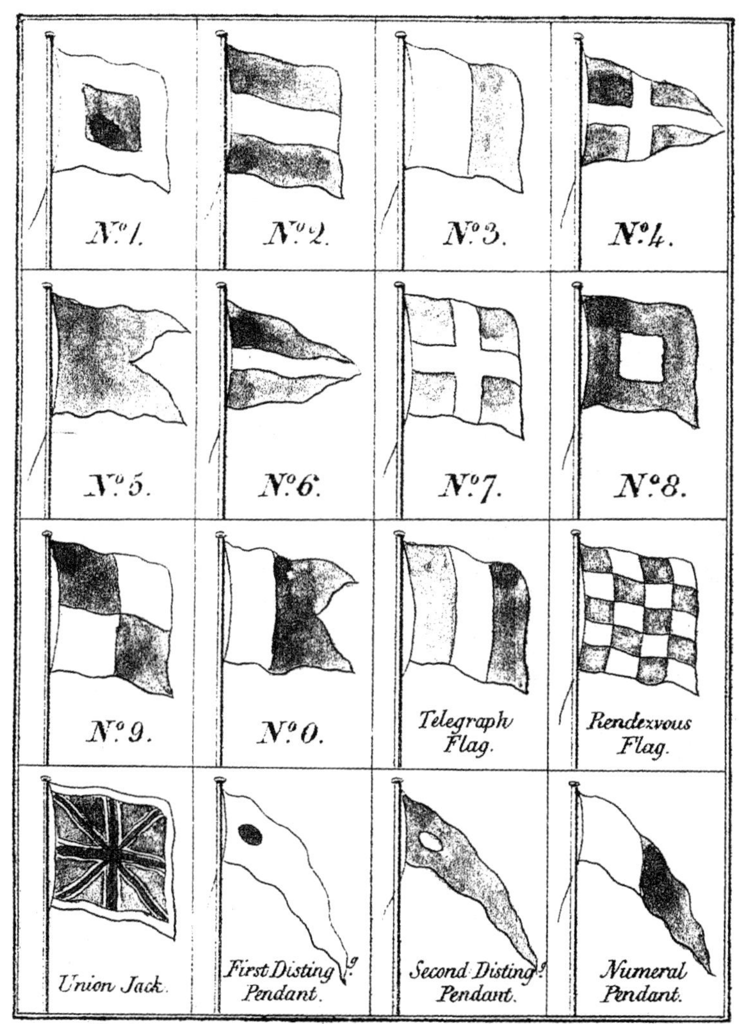

The signal flag system Captain Marryat devised could be understood by warships, merchant vessels and yachts. (Marryat, *A Code of Signals for the Use of Vessels Employed in the Merchant Service*, 1841)

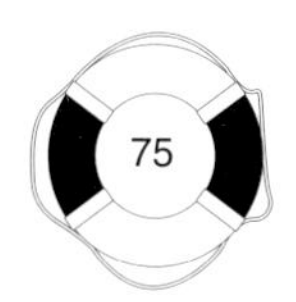

NAVAL GUN SALUTES

RANK	NUMBER OF GUNS
Admiral of the Fleet	19
Admiral	17
Vice-Admiral	15
Rear Admiral	13
Commodore	11

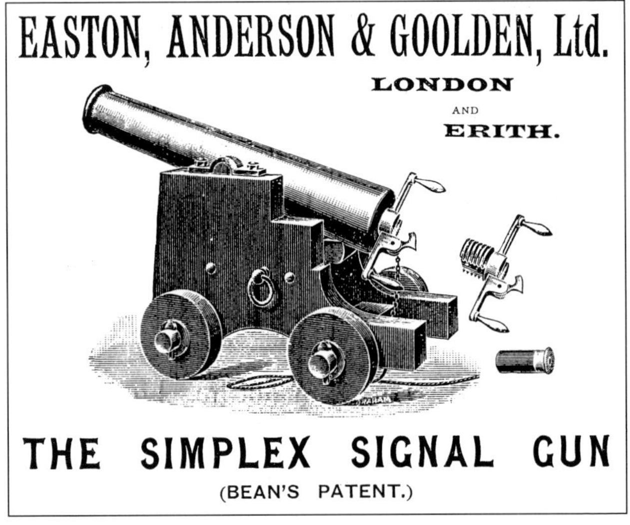

In the 1890s yachts clubs could purchase breach-loading signal guns. (Imperial Institute, *Fisheries and Yachting Exhibition*, 1897)

William Evans's Navigation Sidelights

The idea of using running lights to distinguish vessels at night had been devised in its simplest form in 1420 by Piero Mozenigo, Venice's Captain-General. 'Mozenigo's Code' required that his squadron commanders should show three lanterns at night, and

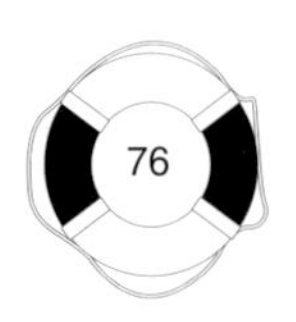

other Venetian ships one lantern. Various maritime nations allowed their sailing vessels to indicate their positions at night by large candle-lit lanterns attached to the poop rail, yet there was no universal requirement for them to do so. However, in the early decades of the nineteenth century the accelerating number of collisions involving the new-fangled steamships meant there was a rapidly growing necessity for a standardised system of recognition – and the invention on the east coast of America of reliable oil lamps (most notably the hurricane lamp in 1826) created the basis for a sophisticated system of navigation lights.

Until 1834 no one had thought of using static coloured lights on ships for the purposes of pilotage and navigation. That year the City of Dublin Steam Company, which operated steamships in the Irish Sea, decided that as an experiment their vessels would show three lights at night: one at the foremast head, and one on each side in front of the paddle-boxes, shining outwards at an angle of 30 degrees, the port light being in deep red. The disadvantage of this system was that the starboard white light was not easily identifiable as such. Then in February 1836 the *Nautical Magazine* reported that it had received news of an idea for a revolutionary three light system from Captain Evans of the packet steamer HMP *Vixen*. What William Evans was proposing for the purposes of preventing collisions was basically an ingenious modification of the City of Dublin Steam Company's arrangement. Instead of having a white light on the starboard side, in Evans's system that light would be *green*.

Evans had served with the Royal Navy during the latter years of the Napoleonic wars, and since then had been with the Admiralty's packet service, whose ships had responsibility for moving parcels and letters between Britain and Ireland. It was doubtless while his command, HMP *Vixen*, was going about her duties that he learnt about the City of Dublin Steam Company's system. What prompted his modification to it was the publication in 1836 of a Royal

Commission report on pilotage and collisions, which called for new forms of identifying lights. Within months of Evans's proposal, the mail ships at Milford – where Evans was based – were steaming with the experimental white, red and green lights.

Oddly, for many years the system was slow to spread beyond the Milford station. Indeed it was not until 1847 that the Admiralty issued instructions that *all* steam packets should display a white masthead light, and red and green bow lights with inboard screens. In the next year the Regulation of Steam Navigation Act was passed requiring all British steam vessels to carry such lights. In 1849 the Evans system was adopted by France and Sweden, then spread throughout the maritime world. It had taken a while for various teething difficulties to be overcome: the introduction of prismatic lenses, for example, solved the problem of horizontally beamed bevelled lenses shining at a downwards angle in the leeward waves when a ship heeled. If a green light shone duller than the red, the green lantern could be enhanced to be more luminous. Only in 1858 were British sailing vessels required to have red and green sidelights.

However, it was not Evans who took the credit for his innovation. Instead that honour was assumed by a Glaswegian civil engineer called Robert Rettie, who attempted to adapt Evans's three colour navigation light system into a signalling system he had devised in 1841. Rettie's idea of sending signals by a combination of hoisted lights and a magic lantern fitted with red and green filters was tested for the Admiralty by HMS *Comet*, a steam vessel, off Spithead in 1845. Eventually the Admiralty decided not to accept it as a signal system. However, such was Rettie's skill as a publicist – he even circulated pamphlets describing the system – that the concept of using white, red and green lights was popularly assumed to have been his.

Having retired from the Navy on reaching fifty in 1840, William Evans went on to achieve fame as a brilliant chess player. As long ago as 1824, while running packets between Milford Haven and

Entered at Stationers Hall.

RETTIE'S

SIGNALS FOR SAILING VESSELS.

For the use of the Patent Signal Lamp for preventing collision at Sea adapted for

RAILWAYS, STEAMBOATS, SAILING VESSELS, &c.

SAILING VESSEL SIGNALS.

N.º 1. INTERROGATION SIGNAL ......
CLEAR LIGHT Top Mast (Time 2 Minutes)

N.º 2. SAILING VESSEL'S SIGNAL or Reply
½ CLEAR ½ GREEN ......

N.º 3. CLEAR ..... alls right

N.º 4. GREEN ..... Starboard

N.º 5. RED ..... Larboard

N.º 6. SIGNAL OF DISTRESS
CLEAR ..... at Top Mast
GREEN ..... at Cross Trees

N.º 7. SIGNAL FOR BOAT OR PILOT
CLEAR LIGHT at Top Mast. 5 Minutes.
If more than one Boat increase the Number of Lights at Top Mast.

N.B. The above signals are laid down for the use of all classes of Sailing Vessels, and from their simplicity and general utility the Inventor requests the attention of Proprietors of Steam or Sailing vessels to give them a trial. – All sizes of Lamps suitable for Railways, Steamboats & Sailing Vessels, or Tickets will be supplied by the Patentee for altering the old Lamps on application at the Inventor. 148, Argyll S.t Glasgow.

Published for & by the Proprietor, Rob.t Rettie, 148 Argyll S.t Glasgow.
4.th June. 1841.

41.
11. 8.
1063c.

The public acclaim Captain William Evans deserved for his three-light system instead went to Robert Rettie, an engineer who in 1841 sought to innovate it. (Rettie, *Universal System of Marine Night Signals*, 1847)

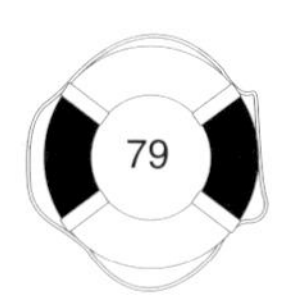

Waterford, he had devised a brilliant chess game opening, which later became known the world over as the 'Evans Gambit'. Two years later he created a sensation when, introducing his opening at an important chess tournament, he defeated the strongest player Ireland had ever produced. He further made history in 1846 by participating in the first ever chess game played via telegraph.

Evans spent his retirement years in London's chess clubs and travelling abroad, sometimes as far as Cape Verde. However, by 1871, when he was aged eighty-one, his savings were all gone and he was virtually destitute. An announcement later that year, that the government was arranging a competition with a £100 prize for the inventor of the most efficient form of navigation lantern, sparked a storm of protest in the letters pages of the *Gentleman's Journal Supplement* in June 1872, complaining that the real creator of navigation lights had been shabbily treated. Shamed into action, the British Government provided Evans with a £1,500 reward, and he also received a £200 donation and a gold chronometer from the Tsar of Russia! Evans's accomplishment as the creator of the tricolour light system might still have gone entirely unnoticed had it not been fleetingly mentioned in a biographical article on him which appeared in the *British Chess Magazine* in 1928. This revealed that the only existing acknowledgement of his nautical innovation

INFAMOUS YACHTING ACCIDENTS

YACHT	SKIPPER	YEAR	CAUSE	CONSEQUENCE
Satanita	A.D. Clarke	1894	Avoiding collision	Sank *Valkyrie II*
Gipsy Moth V	Desmond Hampton	1982	Skipper fell asleep	Wrecked
Drum of England	Simon Le Bon	1985	Keel fell off	Capsized
Team Philips	Pete Goss	2000	Port bow snapped	Unseaworthy
Seamaster	Peter Blake	2001	Boarded by pirates	Blake killed

was the inscription on his tombstone in Ostend, Belgium, where he had died in early August 1872 – within months of the appeal for a financial subscription for him.

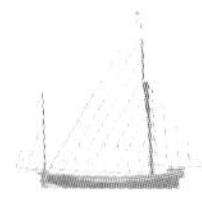

Various types of early navigation lights can be seen at the National Maritime Museum, Greenwich, www.nmm.ac.uk.

Philip Columb's Morse Lamp

Philip Columb might have had an undistinguished naval career had he not, in 1858, been ordered by the Admiralty to find a means of developing a signalling system. After many months' work he devised a scheme which eventually became known as 'Columb's Flashing Signals'. It was, in fact, an application of the Morse telegraphic system, in which the movements of the needle were replaced by long and short flashes from a lamp by night.

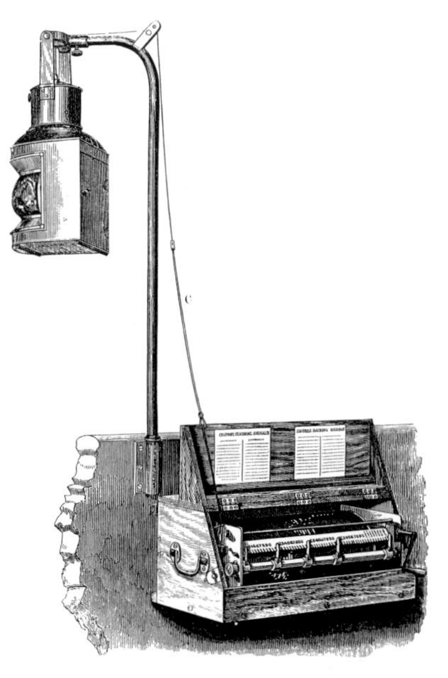

Philip Columb's flashing signals contraption had a mechanism like a music box and could quickly blink preset messages. (Columb and Bolton, *Flashing Signals*, 1870)

The new system was regarded with much scepticism. On 16 July Columb was attached to HMS *Edgar*, the flagship of Admiral Dacres's Channel Fleet, to supervise a programme of experiments, and he was allowed a quarter of an hour to instruct a few signalmen. That night an impromptu signal was so quickly sent, understood and answered that Dacres immediately accepted the system. The apparatus was supplied to every ship in the Channel Fleet and to many in the Mediterranean, and was fully adopted by the Navy in 1867. That year Columb was lent to the Royal Engineers to improve the Army's system of military signalling.

For the Admiralty, Columb also devised a boat-sized version! (Columb and Bolton, *Flashing Signals*, 1870)

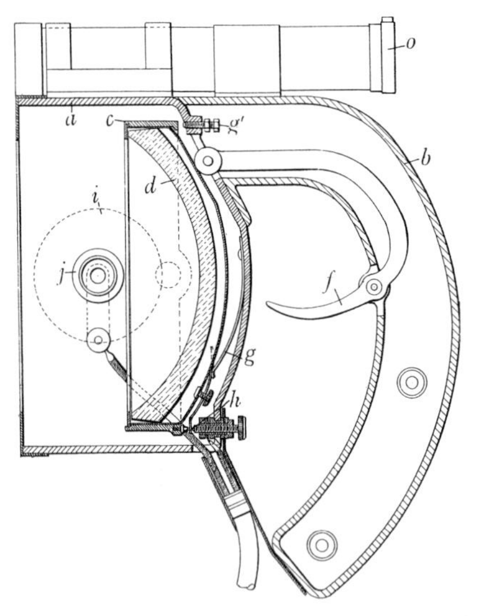

Yachtsmen had to wait until after the First World War for a portable flashing signals device: the Aldis Lamp, invented in 1916 by a Birmingham optician called Arthur Aldis – for signalling from aircraft. (Patent Office, 1916)

The spring shutter machine itself was initially like a musical box, which could be set with the signal pendant numbers of any ship. Only an inexpert rating was needed to turn the handle, and the apparatus would continue to flash the pendant numbers until the ship's attention had been acknowledged. Columb then improved the device by adding a frame with pivoted slats, like a Venetian blind. Many years later such slats were applied to searchlights in the Navy, becoming known as Scott's shutter, having been revived by Admiral Sir Percy Scott, who thus took the credit for an invention which had actually been made many years earlier by Columb. The introduction during the First World War of the portable version of it, the Aldis lamp (invented by Arthur Aldis), enabled light signalling to be done from small naval vessels and eventually from yachts.

There are various naval signalling devices at the National Maritime Museum, Greenwich, www.nmm.ac.uk.

Robert FitzRoy, Weather Forecaster

The survey ship HMS *Beagle* eventually became famous for having on board the influential naturalist Charles Darwin during her epic chart-making and meridian distance checking voyage around the world in 1831–6. Rather less well-known was an incident that involved *Beagle* during a previous hydrographic survey off the Patagonian coast. In February 1829 *Beagle*'s new commanding officer, Robert FitzRoy (her previous captain having committed suicide), had nearly sailed the vessel into sheltered water when a sudden violent squall blew the 240-ton brig on her beam-ends.

FitzRoy's superb seamanship just saved the *Beagle* from turning turtle and sinking, but he could never forgive himself for ignoring the sharp drop in barometric pressure that he had noticed before the squall. The near loss of the ship, and the deaths of two crewmen who were swept from the rigging and drowned, had a profound effect on the religiously conscientious FitzRoy. The traumatic episode set him wondering if a more systematic means could be devised to foretell bad weather, and in 1843 he suggested to a parliamentary select committee on shipwrecks that barometers be distributed along the coast of Britain to provide early warning of

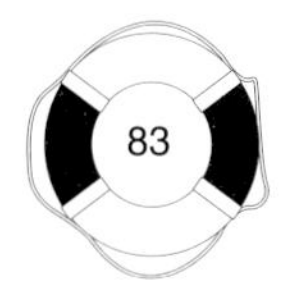

storms. Various career moves prevented him from doing more to encourage such reforms – in 1843–6, for example, he was an outspoken Governor of New Zealand who insisted that Maori land claims should be considered as valid as those of the settlers. The opportunity for further development did not come until 1854 when the Board of Trade decided to create a Meteorological Department, and upon the recommendation of the Royal Society – of which he was a Fellow – selected FitzRoy to be its superintendent because of his concern for weather matters.

By 1857 he was an admiral and had designed the simple robust 'Fishery barometer' (which became known as the FitzRoy barometer). On the apparatus were inscribed weather lore rhymes, such as: 'When rise begins after low, Squalls expect and clear blow' – a verse that FitzRoy himself is believed to have composed on the *Beagle*. Realising that fishermen off the coasts of Britain were liable to suffer most in bad weather, he was particularly concerned about their well-being. With the financial assistance of some eminent philanthropists he arranged to have initially thirteen, and ultimately nearly a hundred, of those barometers manufactured and distributed, complete with instruction booklets, to various seafaring centres and lifeboat stations.

Despite his efforts, in the phenomenally hot summer of 1859 a well-found ship called the *Royal Charter* perished with nearly four hundred souls and a huge cargo of gold bullion in a storm off the Welsh coast. A particularly shocking maritime tragedy, it led to a report by the British Association for the Advancement of Science which recommended that the Meteorological Department should further develop its function of collecting weather statistics, and should use the new telegraph network to send storm warnings to coastal centres – albeit the storms were probably already in progress! Seizing the chance provided by such a widespread readiness for reform, FitzRoy went even further and developed his

Admiral Robert FitzRoy, the inventor of weather forecasts. (Science Museum/Science and Society Picture Library)

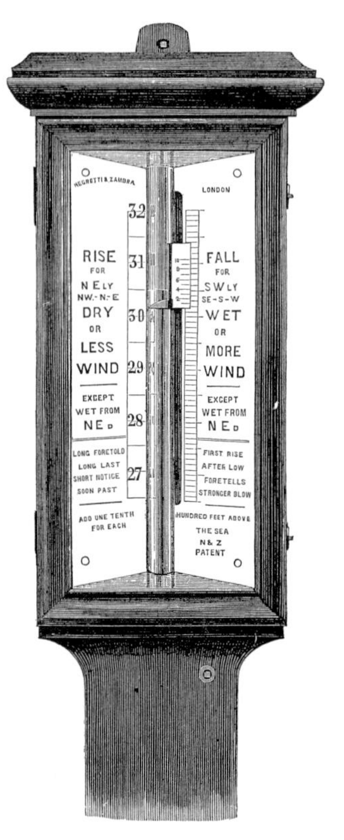

Barometers designed by FitzRoy were publicly viewable at coastal ports. (Lewis, *The Lifeboat and its Work*, 1874)

department into a unit for producing 'weather forecasts' (a term he had invented back in 1855).

As a consequence, as early as February 1861 FitzRoy had in place a package of reforms for administering weather information. Trained agents in various British and Irish ports telegraphed messages about local weather conditions, including observations of clouds, with such communications also coming in from as far afield as Valencia, Copenhagen, Brest and Lisbon. At his London office Admiral FitzRoy would use the data to compile a synoptic chart of changing weather patterns, then he would telegraph any relevant findings back to the local agents. The priority was to detect imminent storms, and if there was any likelihood of one then the local ports and harbours would be instructed to hoist in a prominent location a simple system of warning storm cones (also invented by FitzRoy).

Encouraged by the apparent success of these 'storm warnings', FitzRoy reckoned he could go further still. The telegraph could, he hoped, eventually provide the means for him to produce 10-degree square weather charts for the whole area of the world's oceans. His next step, from August 1861, was to use his synoptic chart to predict the weather of Britain – something no one had ever seriously attempted to do. In 1862 this enabled *The Times* to publish a 'Daily Forecast'.

Mariners and others appreciated the forecasts. Amazingly, an official meteorological inquiry after FitzRoy's death found that some 75 per cent of his predictions and storm warnings were accurate. Yet for newspapers readers they were a novelty and inevitably attracted most attention when they seemed wrong. In 1864 the forecasts were condemned in the House of Commons, and that June even *The Times* ran a leading article ridiculing them.

For FitzRoy it was all becoming too much. The year 1859 had also been a pivotal one for him because of the publication that November of Darwin's bestseller *The Origin of Species*, whose controversial theory of evolution was at variance with FitzRoy's fundamental religious belief in the literal truth of the Bible. Yet the *Beagle*'s

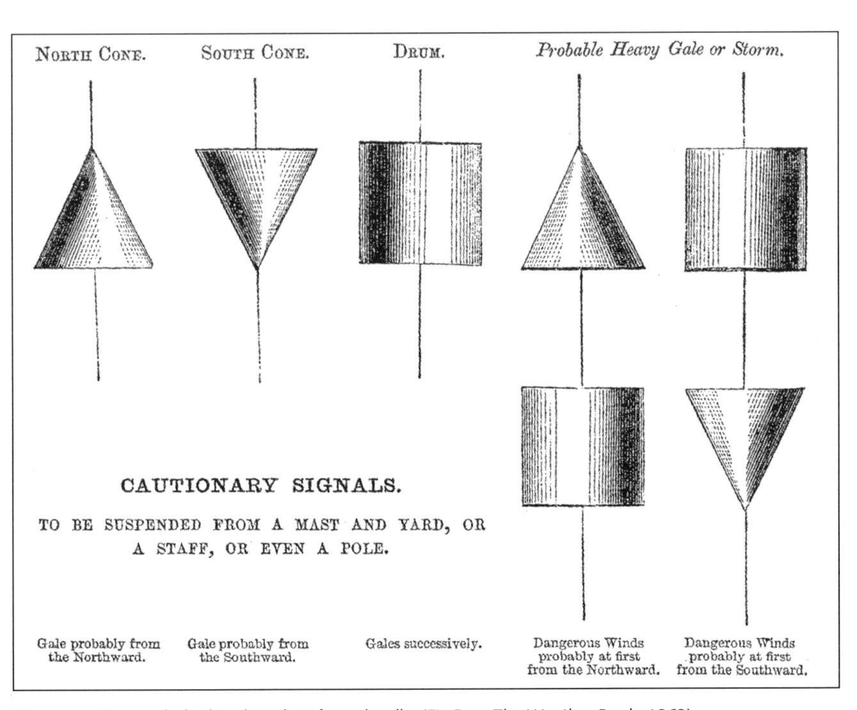

Storm warning symbols also alerted seafarers locally. (FitzRoy, *The Weather Book*, 1862)

association with Darwin meant that the book – a fast growing commercial success – was increasingly linking FitzRoy with these new ideas which were so alien to him. Adding to his depression, FitzRoy's own mighty written work, *The Weather Book*, which appeared in 1862 and ought to have reinforced his place as a leading meteorologist, was being eclipsed by the public criticisms of his forecasts. On top of all this he was working all hours to improve the quality of the forecasts, and was also doing sterling public service as Secretary of the Lifeboat Association. It all ended tragically. Going deaf, and suffering from exhaustion and depression, on 30 April 1865 Admiral FitzRoy committed suicide at his home in Surrey.

FitzRoy had made history in several respects. In command of the *Beagle* he had made her the first vessel in the Royal Navy aboard which the archaic word 'larboard' was replaced with the clearer term 'port', and in his hydrographic activities he had been the innovator of an improved form of surveying quadrant. However, it was in recognition of his services to weather forecasting that in 2002 the Meteorological Office, of which he had been the first superintendant, immortalised his name by renaming the Finisterre sea area Fitzroy.

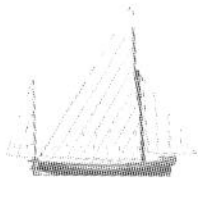

HMS *Beagle* is currently being recovered in the River Roach, Essex. For details of a scheme to build a replica *Beagle*, see www.beagleproject.com.

Lee de Forest's Radio Telephone

Maritime communications today are mostly done by VHF radio, but exactly when a human voice was transmitted by radio for the first time is debatable. Claims to that distinction range from the words reputedly spoken by melon-farmer Nathan Stubblefield to a

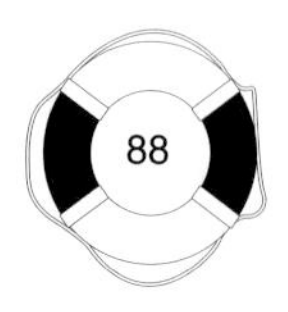

Coastal Weather Forecasts

Better known as a yacht chandler, Captain O.H. Watts was also the creator – and for many years the editor – of *Reed's Nautical Almanac*. A significant consequence of his campaign for better weather information was that in 1965 coastal weather forecasts began to be routinely broadcast on BBC radio. Weather forecasts were first broadcast on television in 1954.

receiving set in Kentucky in 1892 to an experimental programme of talk and music by Reginald Fessenden of Brant Rock, Massachusetts, in 1906.

The pioneer of maritime radio – as opposed to Marconi's voiceless wireless-telegraph system – was a flamboyant entrepreneur called Lee de Forest. A brilliant engineer, who during his lifetime would have nearly two hundred patents to his name, in 1899 he had become the first physics student in America to earn a PhD in wireless technology. A critical component of the early wireless-telegraph sets was the Fleming valve. This was a modification of Thomas Edison's electric lamp by Ambrose Fleming, who had added a second element, but it had a serious flaw in that it could not prevent high and unstable discharges of electricity, which caused interference. However, in 1906 de Forest decided to go a step further and added a grid to the valve to control and amplify signals. He initially hoped that this new three-electrode valve, which he named the audion, might be useful for detecting radio waves. However, he soon found that the valve boosted radio waves as they were received – thus enabling the creation of a new form of communication, 'wireless telephony', in which music and the human voice could be clearly heard.

The audion triggered off a revolution in electrical engineering and made possible radio broadcasting, and eventually even television. It was an invention which could be ranked in importance alongside the invention of printing (a claim magnified by the consummate showman de Forest in his memoirs *Father of Radio*). To publicise its potential, in 1910 de Forest erected a transmitter on the top of the Metropolitan Opera House in New York, and from the stage below broadcast live the voice of Caruso. The US Navy was particularly interested in de Forest's system and employed him to build various transmitting and receiving stations around the Caribbean and to equip a fleet of ships with the de Forest wireless telephones. In 1916 history was made when this apparatus enabled the Secretary of the Navy, Josephus Daniels, to talk directly to a moving ship, the USS *Nebraska*, by radio telephone.

The audion was also used in land-line telephone systems to make long-distance telephone calls affordable. It was left to others to develop the more compact forms of radio telephone, and ultimately VHF sets, which could be used on yachts. De Forest became more interested in cinema technology, and by 1923 had developed a sound-on-film device called the Phonofilm – which enabled him to produce an early form of talking movie.

The Telephone Museum, Wolverton, Milton Keynes, www.mkheritage.co.uk.

The Hammock

The innovator of the hammock was Christopher Columbus, who, on his voyages of discovery, found Caribbean Indians sleeping in nets slung between trees. He adapted the system for his crew and by 1660 the hammock, or 'hamaca' as it was known in Spanish, was widely used in ships. Usually made of extra stout canvas called 'hammocking', the ship-board quality hammocks were found by Victorian yachtsmen not to hold moisture like mattresses, and thus were better for sea work.

4

Sails and Rigging

Nathaniel Butler's Bermuda Rig

The potato was introduced into North America by an adventurous British sea captain, who was also the earliest facilitator of the development of the Bermuda rig. Since being brought to Europe by adventurers such as Sir Walter Raleigh, the potato had become popularised and ready to be returned by colonists westwards across the Atlantic. Cedar chests containing the potatoes and other Christmas goodies for the Governor of Virginia, Francis Wyatt, reached the mainland in December 1621, having been sent as goodwill presents by Nathaniel Butler (sometimes Boteler), Governor of Bermuda. Despite his rank, he had spent much of that year attempting to seize salvage from the wreck of the *San Antonio*, a treasure ship which had foundered off the island. Bermuda had only been discovered by British seafarers in 1609 and Butler had been appointed Governor of the fledgling colony in 1619, since when he had been determined to get his hands on such plunder and for that purpose had ordered two cedar boats to be built. By chance there had happened to be in Bermuda a skilled Dutch carpenter who had recently been marooned there when the frigate on which he had served was wrecked. So reputedly excellent were the Dutchman's sailing craft that he made a few more. Local people clamoured to buy them, but soon complaints were finding their way to London. Although the buccaneering Governor was claiming that the carpenter was retained on Bermuda for the purposes of making boats for the public, it seems that Butler had actually begun selling the craft for his own profit.

A woodcut illustration of a vessel showing the basic form of the Dutch carpenter's rig appeared in a 1671 book of poems by John Hardie describing a voyage to Bermuda. It consisted of two tall jib-headed sails, without booms, each laced to its own mast. There was no jib, the foremast was stepped as far forward as it would go, and both masts were without shrouds or stays, and were very whippy. They raked so severely that each sail was an isosceles triangle. The design was heavily influenced by the weather and water conditions in Bermuda, as the steady winds permitted the use of large rigs

Seventeenth-century woodcut depicting a Bermuda rig boat in Nathaniel Butler's time. (Lefroy, *Memorials of the Bermudas*, 1877)

while the many narrow channels between the islands put a premium on windward ability. Such a rig was as yet unheard of in England, but even in the early seventeenth century it was not uncommon in the Netherlands and it was natural that the carpenter would have built and rigged boats after the manner of his own country. The idea of a lofty triangular sail may well have been derived by the Dutch from the lateen sails of southern Europe, which they had apparently developed from an Arab-style diagonally hung spar rig by converting the spar into a mast.

The earliest accounts to circulate in Britain about the Bermuda rig's sailing qualities were all favourable. In 1675 Samuel Fortrey, a merchant of Dutch ancestry who lived at the famous Dutch House by the Thames at Kew, went so far as to produce a treatise called *Of Navarchi*. Although he sent it to Samuel Pepys (in his capacity as an Admiralty personage), suggesting how the windward performance of a sail such as the 'Bermoodn' could be enhanced by the addition of a 'Boome', Pepys apparently never responded; and it has never been established quite how, or when, the boom and jib eventually came to be standard features of the Bermuda rig.

So far only two-masted schooners in the islands carried the distinctive Bermuda rig. The yachtsman who made the next quantum leap in the rig's evolution was the Honourable H.G. Hunt, who, along with fellow British naval officers stationed in Bermuda in the 1820s, keenly raced in local boats there. His small schooner having been defeated, Hunt wondered if a one-masted boat might be faster, particularly to windward. Forthwith he had a sloop built and challenged his rivals to another contest. Meanwhile he secretly discovered the superiority of that single mainsail by means of a private race with a schooner at midnight, and then on the following day he was convincingly victorious in a public sailing match. His success in both races drew the attention of his brother officers to the effectiveness of his new form of sloop. Hitherto the

A nineteenth-century Bermudan yacht showing its influential rig. (Folkard, *The Sailing Boat*, 1906)

schooner rig had been ubiquitous in the islands, but from now on the preferred arrangement became that with one mast only. As the new variety of rig developed it received wider attention, some British officers having taken their boats home with them to England. A few of these were reputedly raced against English yachts – and won with ease.

Yet it was in America – where it was referred to as 'jib-headed' – that the Bermuda rig really caught on, most notably with two-masted commercial vessels such as the Chesapeake Bay sharpies and bugeyes. It was not significantly used for yacht racing until the William Pickard Stephens-designed *Ethelwynn*, which had a waterline length of just 15 feet, decisively defeated her gaff-rigged rivals and successfully defended the first Seawanhaka International Trophy in 1896.

In Britain there was scepticism about the seaworthiness of the Bermuda rig. It was believed – without much substantiating evidence – that the yachts brought back from the islands in about 1834 were unsuited to the stormy weather which boats in Britain sometimes faced. The fact that the Bermudan boats had unstayed masts saw them dismissed as fit only to sail in light breezes. In the 1863 third edition of his book *The Sailing Boat*, the respected yachting commentator Henry Folkard cautioned: 'The Bermudan rig ought not to be extended to vessels above eighteen tons, because any mast fit to carry a proportionate area of canvas would be too ponderous for its position.' However, by doing away with the heavy conventional gaff the Bermudan rig could use a lighter, though necessarily longer, mast. Such an evolution was facilitated by the introduction of hollow masts and strong, virtually waterproof glues. Improvements were made to make the rigs sturdier; for example, instead of the Bermuda-style attachment of the boom to the mast by means of an eyelet, it was now pivoted with a gooseneck. This saw the rig develop into the Anglo-Bermudan rig.

In 1912 the 15-metre cutter *Istria* made history by becoming the first significant British yacht in the twentieth century to step a mainmast and topmast in one spar. So many wires, diamond shrouds, spreaders and jumper struts were required to cope with such tall masts that the results resembled early wireless towers, and unimpressed traditionalists dubbed the new-fangled Bermudan rigging the 'Marconi' rig. They remained doubtful as to whether the Bermudan rig could perform better than the standard working gaff rig. 'To our mind', Frank Cowper declared in his 1890 cruising book *Sailing Tours*, 'there is nothing so handy as a cutter, or rather a modification of the cutter. For small boats, the balanced lugsail with a jib is the perfection of rig; they are very fast and are as handy as possible.' Such views

The Highfield Lever

This hand-operated lever, used for slackening off or tautening running backstays and forestays, was invented in 1930 by a commodore of the Royal Thames Yacht Club called Somerville Highfield. Tested aboard his 15-metre racing yacht *Dorina*, the patent Highfield Lever provided the newly introduced large Bermuda rigs with a faster means of tensioning control than the old-fashioned block-and-tackle method had done. *Dorina* was subsequently sold by Highfield. Renamed *Kismett III*, she became the first yacht to finish the 1935 Fastnet race.

were eventually scientifically disproved in 1915 by an aerodynamicist called Professor George Owen, who experimented with the modern rig aboard an R-boat at Marblehead, then tested model yachts in a wind-tunnel at the Massachusetts Institute of Technology. Owen found the Bermuda sail performed best to windward and could develop 20 per cent more driving force on a broad reach.

For the 1920 season the huge 23-metre *Nyria*, owned by Mrs R.E. Workman, was re-rigged in Bermudan style; then in 1928 appeared the William Fife-designed *Hallowe'en*, the first British cruising yacht with a Bermudan mainsail. Controversy about the rig was finally ended in 1931 when King George V's *Britannia* was re-rigged as a Bermuda cutter – stepping a hollow spruce mast 175ft tall! The rig had taken all of 310 years to become established since the innovative work of Governor Nathaniel Butler, the influential potato enthusiast.

Nathaniel Butler's official residence, The Old State House, is Bermuda's oldest stone dwelling, and a traditional Bermudan dinghy is exhibited at the Bermuda Museum, www.bermuda-online.org.

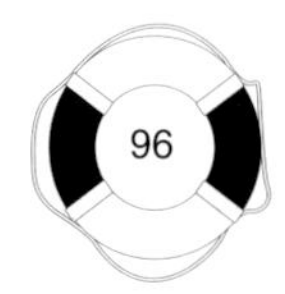

William Gordon's Spinnaker

Booming out foresails by the clew when running had for countless decades, until the mid-nineteenth century, been a common practice on fishing smacks. It was known that for many craft running was the fastest point of sailing, but whereas the working boats' traditional yard-hung square mainsails had peripheral drabblers and bonnets at their edges, a good-sized running canvas had yet to be devised for yachts. For leisure boating there had been little incentive to do so because the most influential racing clubs, such as the Royal Thames, disallowed extraneous headsails and the ungentlemanly practice of booming out. Unwisely, in 1851 the booming-out restriction had been waived for a visiting schooner called *America* – which then won the first America's Cup race around the Isle of Wight.

Until the spinnaker was invented in 1866, yachts boomed out square sails (and sometimes jibs), with a conventional yardarm. (Dixon Kemp, *A Manual of Yacht and Boat Sailing*, 1891)

All we know about the Hampshire yachtsman William Gordon who invented a radical new sail for yachts in 1865 is that he was a prosperous Southampton sailmaker who owned a yacht called *Niobe*, designed locally by Hatcher of Southampton. William Gordon evidently had a determined entrepreneurial spirit and was ready to capitalise on such opportunities as came his way. In the summer of 1865, for example, the Royal London club rescinded its ban on booming out, and on 5 June, during a run in a Royal London match, *Niobe* hoisted to her topmast head a large triangular sail which was allowed to boom out freely, enabling her to draw away rapidly from her rivals. The considerable attention this aroused meant that for a while the sail was known as a 'Ni-ohs' or a 'Niobe'.

Perhaps William Gordon expected his new sail to end up bearing his name – but once more this was not to be. In 1866 he sold one of the several 'Ni-ohs' his Southampton sail-loft had quickly manufactured to a Mr H.C. Maudslay, the owner of the 48-ton cutter *Sphinx*. On 15 August she took part in a race organised by the Royal Victorian Yacht Club at Ryde pier around the Nab lightship, but, instead of letting her Niobe blow out freely, *Sphinx* bent hers to a boom which brought the clew further out to leeward and made the sail set better. This enabled *Sphinx* to gain second place in the event and a prize of £25!

Dixon Kemp, the yachting editor of *The Field*, happened to be on board *Sphinx* at Maudslay's invitation and overheard the crew discussing what to call the new-fangled piece of running canvas. *Sphinx*'s deckhands and the Itchen ferrymen, who called the recently built yacht the 'Spinx' (or 'Spink'), named it the 'Spinxer'. When one of them quipped 'Now there is the sail to make her spin', one of the afterguard was said to have coined the phrase 'spin-maker'. Dixon Kemp's report of the historic 'Spinniker' event appeared in *The Field* on

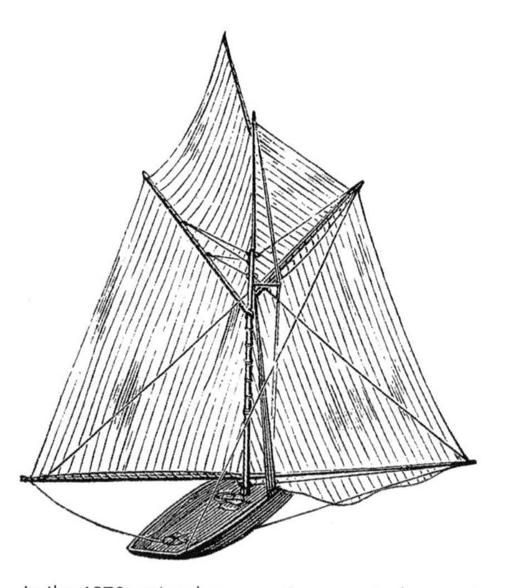

In the 1870s spinnakers sometimes required support from a spar goosenecked to the foreside of the masthead. (Dixon Kemp, *Manual of Yacht and Boat Sailing*, 1891)

OLDEST INTERNATIONAL YACHT CLUBS

YEAR	CLUB	COUNTRY
1718	Flotilla of the Neva	Russia
1720	Water Club of Cork	Ireland
1775	Cumberland Fleet	England
1826	Singapore YC	Singapore
1830	Royal Swedish YC	Sweden

18 August, and the word next appeared in print in the September 1866 edition of *Hunt's Yachting Magazine*.

The sail's name evolved, perhaps by means of rapid pronunciation, from 'Spinniker' into 'Spinnaker' and by 1869 that form of spelling had become established. Subsequently all manner of balloon sails appeared, ranging from the fairly flat spinnakers used by the 1881 America's Cup challenger *Atalanta* and defender *Mischief*, to the bulging J-class spinnakers of the 1930s with experimental round holes. In 1927 the symmetrical 'double spinnaker' appeared, which was the easiest to gybe. There were also several exotic varieties of cut, such as the precariously bulging 'Herbulot'; the tri-radical spinnaker originally developed by Ken Rose of Banks Sails; and the multi-coloured triangular New Zealand 'Blooper' spinnaker, which was eventually adapted into a safe cruising chute.

While the hapless William Gordon vanished into obscurity, Maudslay did rather better. His full name was Herbert Charles Maudslay and he was a descendent of Henry Maudslay (1771–1831), the famous inventor of the labour-saving naval block-cutting contraption. An heir to Henry Maudslay's shipbuilding empire, he continued to be a keen supporter of yachting in the Solent, where he became an honorary secretary of the Sea View Yacht Club.

Sea View Yacht Club, www.svyc.co.uk.

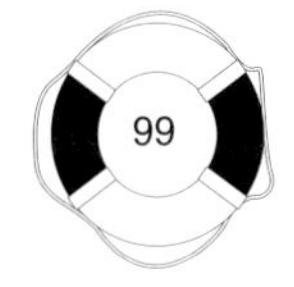

Robert Wykeham-Martin's Furling Jib

The earliest mechanical system for furling sails was patented in 1850 by Henry Cunningham RN as a means of reefing square riggers' topsails remotely from the deck. It ought to have spared crew members the hazards of working yardarms aloft, and allowed such alterations to be done more quickly than by traditional means, and with fewer personnel, but in practice the system was so unreliable owing to frequent jamming that few ships ever installed it. By the 1870s an inventor called Parker March had devised a roller-reefing boom which could be rotated by a simple worm-gear and handle at the gooseneck. Made by various chandlery manufacturers, such as Woodnutt on the Isle of Wight, it became

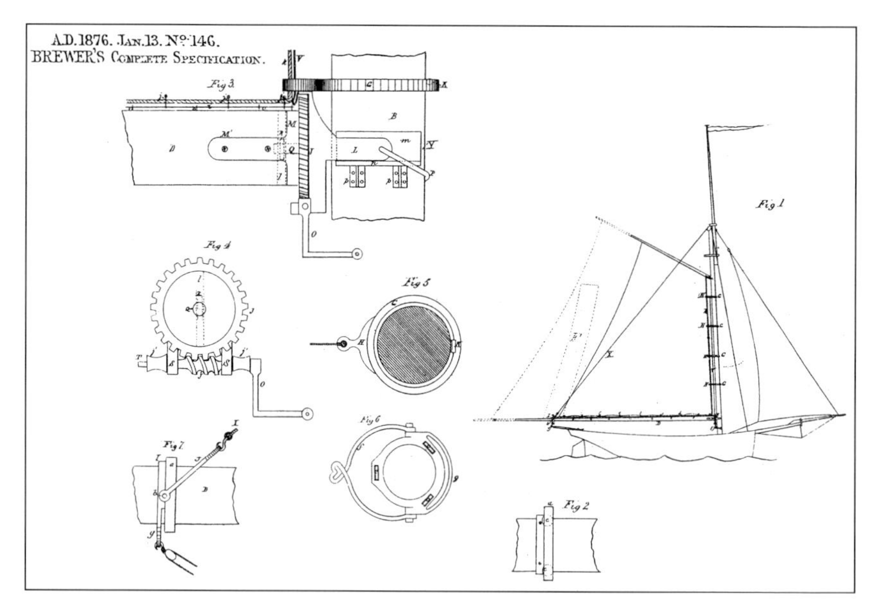

The roller-reefing boom, originally invented by Parker Charles March of Massachusetts, was patented in Britain in 1876. (Patent Office, 1876)

known as the Appledore gear – and it was famously ideal for Bristol Channel pilot cutters which often needed to be sailed short-handed. Other suppliers, like Turners of Beccles, made their own variations using a ratchet mechanism. The process of development seemed to be complete when Lieutenant H.S. Tipping invented a mainsheet strop, an ingenious claw-like device that would secure the upper mainsheet block to the boom even when the mainsail was reefed.

In the 1890s Ernest du Boulay, a Royal Artillery major who sailed at Bembridge, had the brilliant idea of slanting such a roller diagonally for the purposes of jib furling. This seemed likely to become a roaring success. Du Boulay's system had the luff of a jib fastened to a 1-inch diameter pine tube which was rotated around a headstay by a cord pulling on a reel secured to the stemhead. It was in effect an elongated version of the type of roller window-blind found in many Victorian houses. Unfortunately in practice it discoloured and rotted sails, and even pulled them out of shape. Worse still, the damp hollow tube so greatly accelerated wire corrosion that within two or three seasons there were numerous incidents of the entire apparatus disintegrating, sometimes even bringing down the mainmast with it!

When such a roller came apart on his boat, the amateur yachtsman Robert Wykeham-Martin began to wonder how the system could be improved. A poor distant cousin of the wealthy Wykeham-Martins who owned the picturesque Leeds Castle in Kent, Robert lived at Bourne End on the Thames – where he did much of his sailing – and ran a small electrical engineering business called Moody's near St James's Park, Westminster. An innovative character, he would eventually become the inventor of a 'flat-pack' folding yacht tender and a patent flexible bilge pump for use in dinghies. Realising it was the roller itself that was the problem, Wykeham-Martin devised a simple alternative which discarded the wooden tube altogether, thereby reducing weight and windage aloft. Instead he used a flanged drum attached to the stem plate, with a

Jib furlers were ideal for sailing cruisers, such as this on the Norfolk Broads. (Author)

swivel secured to the mast at the head of the foresail; between them ran a standard luff rope sewn to the jib and, if the foresail required hanks, a wire closely parallel to that. It all meant that the sail could be furled around its own luff rope. Because no roller was necessary, the entire contraption could be lowered, coiled and stowed below, and another different-sized jib clipped on and hoisted in minutes.

The Wykeham-Martin furling system was patented and went into production in 1907, and Robert would often insist on assembling the swivelling parts himself to ensure reliability. In his workshop he would pack the bronze bearings with grease and fit them with the ballbearings he bought in bulk at a bicycle shop! Available direct from Moody's, customers could purchase four different sizes. The largest was suitable for large sail areas of up 230sq ft, while the basic model was appropriate for sail areas of no more than 40 sq. feet – prices from £1! Several yachting writers enthused about the practicalities of the Wykeham-Martin furling gear. Francis Cooke, for example, praised it in *The Corinthian Yachtsman's Handbook* (1913) as 'one of the most handy contrivances I have ever used on a yacht'. In response to criticisms that the Wykeham-Martin device was not a reefing gear but only a *furling* gear – because there was nothing to prevent the head from unrolling when furled – Robert developed the system further by replacing the head swivel with a flanged drum and rolling line, with a movable counterweight on one of the shrouds in the form of an old bit of lead piping!

Although it has been continuously in production for nearly a hundred years, the Wykeham-Martin furler has always tended to be more at home on unglamorous cruising yachts, such as those to be found on the Norfolk Broads. Perhaps ironically, since the 1960s there has been a revival of the du Boulay type of reefer, now with a

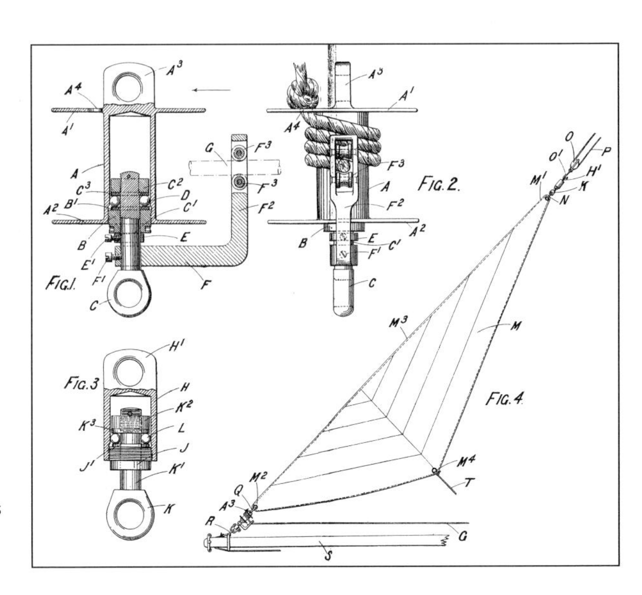

Robert Wykeham-Martin's furling gear patent. (Patent Office, 1907)

metal roller tube, although this too has now also come to be commonly referred to as Wykeham-Martin gear.

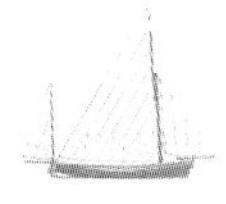

Examples of the Appledore reefing gear can be seen at the North Devon Maritime Museum, Appledore, www.devonmuseums.net.

Sven Salén's Genoa

In 1925 a book was published called *Yacht Racing: The Aerodynamics of Sails* which had an unexpectedly far-reaching influence on the shape of sails. Its author was Manfred Curry, a German aerodynamicist who happened to be one of the best dinghy-racing skippers in Europe. What made the book so popular was that never before had such institutionally gathered scientific data on yacht-sail performance been made public. After years of wind-tunnel research using model sails at the Junkers aircraft works at Dessau, Curry's book revealed that a huge jib could be particularly effective if it also ran back alongside a yacht's mainsail. The book was initially only published in Germany and was mostly concerned with racing tactics, so perhaps it was hardly surprising that most yachtsmen missed its really sensational section which was on experimental sail forms.

The Clamcleat

In 1963 the accidental parting of an electric light pull-cord from its bobble triggered off a thought in inventor Reginald Emery's mind that there must be a better means of attachment. He thus created the Clamcleat. It consisted of two pieces of perspex, in each of which ridges were machined at an angle for increased friction. Scepticism greeted this simple device – the first innovation in rope cleats for over forty years – but the first nylon Clamcleat went into production in 1964.

The first sailing enthusiast to recognise the potential of a large jib sail was the Stockholm-based shipping tycoon Sven Salén, who dared to experiment with it in the International 6-metre class. Originally created in 1906, by the 1920s the Olympic class 6-metre yachts, or 'Sixes', were in their heyday, having established themselves at the highest level of international keel-boat racing. For the 1927 International Regatta, held that February in Genoa, Salén arrived to compete in his 6-metre yacht *Lilian*. He opted to test out Curry's radical theory by the evolutionary step of cutting flat the conventional balloon staysail, which all the Sixes used for light reaching. Hitherto *Lilian* had never been especially fast, nor was she during the reaching phase of the first race at Genoa. But arriving at the leeward mark, at which point all the Sixes usually lowered their staysails, to the astonishment of the other competitors Salén just hardened his in. *Lilian* duly forged ahead with her flat staysail and won the 12-mile event by more than 3 minutes – defying the unwritten law of boat physics that a working jib was best to windward.

Dubbed by onlookers the 'overlapping jib', or 'Swedish jib' (probably because of the nationality of its inventor), the significance of the new wonder sail was initially not universally acknowledged. In July 1927 Salén used it to devastating effect in another Six, *May-be*, to gain the best overall results at the Copenhagen International Regatta. But even then the premier racing magazine *Yachting World* commented only on the construction quality of *May-be*, the new Estlander-designed boat he had been using, and his perfect helming. Thus those hoping to emulate Salén's achievement had little to go on. Nevertheless, later that summer a pioneering yachtsman called Sherman Hoyt from City Island, New York, procured a 6-metre yacht with a mast abaft amidships. Principally built with the intention of qualifying for a larger spinnaker, *Atrocia* also had a huge genoa. Sadly this freak-rigged boat lacked speed in its races. Hoyt was a genuis at sailing in light airs, but even he had not quite understood that positioning was

In 1927 an eccentric alternative to Sven Salén's all-conquering new genoa was the large headsail Sherman Hoyt had on the centre-masted 6-metre *Atrocia*, with which Hoyt hoped to achieve some advantage downwind that season. (Ratsey and de Fontaine, *Yacht Sails*, 1948)

more crucial than size to a genoa's effectiveness. In contrast, Salén had realised from Dr Curry's writings that in order to obtain the optimum 'slot effect' – whereby a foresail increases the power of a mainsail by compressing, and therefore accelerating, the wind through the narrowing 'slot' between the foresail and the lee side of the mainsail – the genoa needed to have a substantial overlap to smooth out that leeward flow.

Curiously, it was with the public acclaim that another of Sven Salén's ingenious sail creations received that the genoa eventually achieved the wider attention it deserved. The Six event off Oyster Bay in Long Island Sound for the Scandinavian Gold Cup series in September 1927 drew the largest number of yachts from various countries that had ever raced together at one time in the USA. At this exciting international event there appeared for the first time a revolutionary parachute-shaped 'double spinnaker' – invented by Sven Salén! It was also the first time the genoa had been put through its paces in America, and that signified a turning point in American yachting history. *May-be*'s success in winning the Gold Cup that year against stiff competition ensured widespread recognition of the radical foresail as a winning device. At Salén's insistence, from then on the sail was known not as the yankee jib but as the genoa.

Salén's wondrous staysail quickly caught on and became so popular for both competitive and leisure sailing that it seemed likely most contemporary classes would adopt it. But for J-class yachts, in which the genoa would have been at its most spectacular, the enormous sail was calculated to be beyond the capacity of any crew to handle in all but the mildest weather. That had been a concern for Charles Nicholson in 1934, when he was designing the America's Cup challenger *Endeavour*, until her owner Tommy Sopwith suggested a genoa with a large area of its clew corner missing. Unfortunately for Sopwith, while that radical quadrilateral sail was being tested in the Solent just before *Endeavour* set off to cross the Atlantic, a quick glimpse of it was caught by Sherman Hoyt, who happened to be sailing there. Realising it was something important and new, he made a hasty sketch of it which he sent off post-haste. Within 24 hours of the sketch reaching New York the Ratsey & Lapthorn sail-loft there was at work on an American version of the double-clewed sail. This became known as the 'Greta Garbo' and it eventually helped *Rainbow*, the Cup defender, win the series.

Sven Salén went on to win a bronze medal for Sweden in the 6-metre class at the 1936 Olympics. In business too he excelled, becoming a pioneer of refrigerated merchant shipping and developing his company into Sweden's biggest shipowner. And yet,

Biggest Sails and Spars

Genoa	19,730 sq.ft	*Mirabella V*	2003
Spinnaker	18,030 sq.ft	*Ranger*	1937
Mainsail	16,760 sq.ft	*Mirabella V*	2003
Mast	290ft	*Mirabella V*	2003
Boom	92ft	*Satanita*	1893

despite all his success, he was no yachting elitist. In 1940 he used his knack of exciting leadership to persuade the Scandinavian Sailing Association to organise a competition to design an affordable version of the Dragon and the Six. The competition resulted in an immensely influential form of cruiser-racer called the Folkboat, which in Swedish meant the 'people's boat'.

Starling Burgess's Alloy Mast

A flurry of activity ensued in the spring of 1929 when Harold Vanderbilt of the New York Yacht Club learned of Sir Thomas Lipton's intention to challenge for the America's Cup. At the previous encounter in 1920, Lipton's elderly *Shamrock IV* had rattled the complacent American yachting establishment by winning two races, providing the closest contest so far in the Cup's history. Aware that the eminent British designer Charles Nicholson was already creating *Shamrock V*, a yacht that might well make Lipton the winner, the Americans' anxieties were now compounded by the introduction of rule changes for the America's Cup. The outcome would no longer be determined by time allowances; the course was being shifted from the crowded waters of New York to just off Rhode Island; and the Cup would now involve seven races, not five. In addition, the 1930 series would also be the first to be contested with giant Bermuda-rigged yachts, all of which had to conform to the Universal Rule's J-class.

Just to qualify for the Cup's defence Vanderbilt would need a boat able to beat off three other American contenders. But he had certain advantages. Not only was Vanderbilt a superb helmsman – and supremely wealthy (he had inherited a vast railway fortune), he was blessed with fine organisational skills and a brilliant analytical mind that served him well in business. He also had a passion for gambling, and reputedly invented Contract Bridge. Even so, Vanderbilt knew he was taking quite a risk in hiring Starling Burgess to design his Cup contender, *Enterprise*.

In the 1880s Starling's father, the distinguished naval architect Edward Burgess, had received huge public acclaim for designing three victorious America's Cup defenders (*Puritan*, *Mayflower* and *Volunteer*), but financially he never did well from his accomplishments and in 1891 died from typhoid fever and overwork, aged just forty-three. A national tribute, put in trust for the education of his children, William Starling Burgess and Charles Paine Burgess, ought to have supplied the means for Starling to follow his father's footsteps, but Starling had other ideas. He read English at Harvard and initially devoted his natural ability for design to writing poems and even, it is reputed, inventing the Times Roman typeface! Dabbling in aviation, he invented a seaplane and built the Wright brothers' planes under licence, before opening a plant at Marblehead to produce Burgess-Dunne flying boats for the US Navy. Only belatedly did he decide to concentrate his attention on yacht design – to sensational effect. His 59ft schooner *Nina* was an easy winner of the 1928 Transatlantic Race. Reputedly the first yacht specifically built for ocean racing, she became the first Bermuda-rig racing yacht to cross the Atlantic. However, Starling had yet to prove he could produce a really effective inshore racing yacht.

Burgess decided to design *Enterprise* with a V-bow to make her faster in light airs, and a greater beam for more stability. In keeping with Vanderbilt's determination that no expense be spared to create the most scientifically perfect racing yacht, he experimented with 15ft model hulls at the Naval Model Basin in Washington. This all resulted in the 121ft *Enterprise* being lightly made of thin Tobin bronze plate on nickel-steel frames. To minimise windage her deck was virtually clear, with just a few foot-rails and no skylights, and most of the professional crew were required to be stationed below to

Starling Burgess was an accomplished aircraft builder before he began designing yachts. (© Mystic Seaport, Mystic, CT)

operate twenty-four winches and coiling drums. The winches were, in fact, mostly donated from *Reliance* and *Resolute*, which had used them to defend the Cup in 1903 and 1920 respectively. Nevertheless, it was *Enterprise* that became known as the 'mechanical boat'.

Figuring that for every bit of weight saved aloft 5 times as much ballast could be dispensed with, Burgess was keen to fit *Enterprise* with the lightest possible mast. In consultation with his brother Charles, a noted mathematician and civilian constructor of dirigibles for the US Navy, Starling did some sail and rigging tests at the University of New York's wind-tunnel and then made the radical decision to make the mast of duralumin alloy. Metal masts were not an entirely new idea: Richard Trevithick had obtained a patent for one as long ago as 1809, and even the America's Cup defender *Columbia* had used a steel mast in 1899. But an aluminium-type mast had never been built before. Weighing only 4,000lb, *Enterprise*'s mast would measure 18 inches wide at the deck, tapering to 8 inches aloft at the head (a wooden mast, in contrast, could be expected to weigh 5,000lb, and be 27 inches at the deck and 10 inches aloft). It was created by the aeroplane builder Glenn L. Martin by fastening together innumerable long strips of duralumin with some 80,000 rivets. At 163ft long, it was too bulky to be moved by road or rail, and so was placed on a special barge and floated from the aircraft factory near Baltimore all the way to City Island, New York.

Another innovation was *Enterprise*'s so-called 'Park Avenue' boom. Its inspiration was an article by Manfred Curry, in which the German aerodynamicist had written of the performance advantage to be gained by exaggerating the natural curve at the foot of a mainsail to increase the vacuum on its lee side. Curry had advocated a bending boom, but the Burgess brothers had the brilliant idea of facilitating such movement by means of tracks on transverse slides arranged 18 inches apart across the flat top of the boom. The boom itself was triangular in section and tapered at both ends so it was boat-shaped. It attained a maximum width of 4 feet, and since it appeared to be as wide as some pavements it became known as the 'Park Avenue' boom. Intent on using 19-strand wire for the rigging because it had a smaller diameter than standard wire and this offered a saving in windage, Burgess next helped invent an effective means of gripping the hitherto virtually unspliceable wire. With the assistance of the American Cable Company, he devised 'Tru-loc' steel sleeves, which duly proved to be totally reliable. The new-fangled duralumin mast itself – the tallest single alloy pole ever stepped in a yacht – needed to be

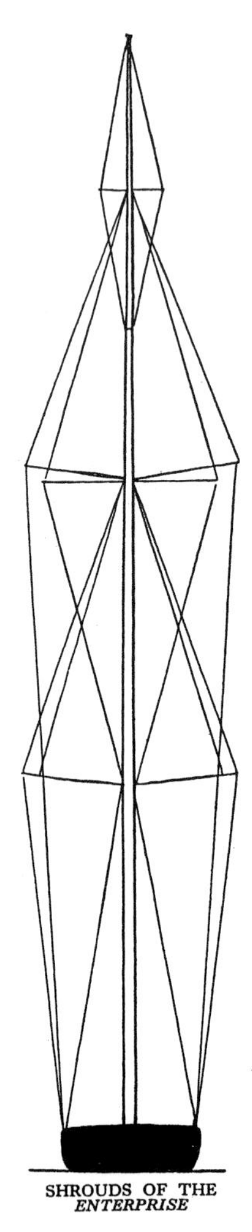

Enterprise's 163ft aluminium mast, devised by Starling Burgess, enabled her successfully to defend the America's Cup in 1930. (Rogers, *Freak Ships*, 1936)

supported by a network of outriggers and truss-bracing wires that were essential for keeping the mast straight. Installed below to monitor the strains on such a fragile rig were various dynamometers and other measuring devices. Overseeing these was Burgess himself, whom Vanderbilt had made responsible for looking after the mast and trimming the rig every time *Enterprise* went about.

Cautiously, Vanderbilt tried out *Enterprise* with a conventional hollow wooden mast. But when this was exchanged for the duralumin one her performance was transformed. She felt like a different boat, lifting more readily over the waves and pounding less because of her improved stability. Starling's various innovations meant she was the lightest of the four defence contenders. In the elimination trials in the summer of 1930 *Enterprise* overcame *Yankee* and *Whirlwind*, but struggled against the Clinton Crane-designed *Weetamore*. Indeed, she was on the brink of failing to be selected, and during the very last elimination race disaster nearly struck. In heavy conditions a dangerously moving spreader threatened *Enterprise*'s mast with imminent collapse. Burgess, with a flash of brilliant improvisation, hooked the ends of the spinnaker halyards together and hauled them in to hold the spreader in place – and it worked. *Enterprise* went on to win and was duly selected as the Cup's defender.

Shamrock V and *Enterprise* eventually met off Rhode Island later that year. The two yachts had similar hull profiles but differed significantly in rig. But *Shamrock V* was under-winched and she had little chance against *Enterprise*'s advantages of performing well to windward, and her innovative boom and lighter rigging. Ironically there was only one rigging failure – a parted halyard cost *Shamrock V* the third race. In the final deciding race, just before the finish, Vanderbilt gave Starling Burgess the honour of steering *Enterprise* to victory. Although *Enterprise* had won all her races crushingly, and thus retained the Cup, it had been at enormous expense. The price of building her, with all her innovations, was estimated then to have been some £150,000 (some four times the cost of *Shamrock V*). It would not be until 1948, when an extrusion process was developed to make aluminium masts, that alloy masts would at last become affordable for less exotic yachts.

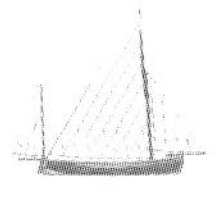

Although *Enterprise* was broken up soon after her epic 1930 victory, *Shamrock V* is afloat and can be chartered, www.rnryachts.com; see also www.jclassyachts.com.

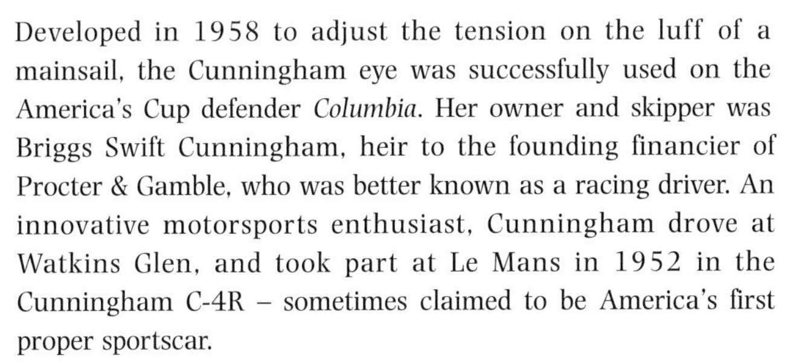

The Cunningham Eye

Developed in 1958 to adjust the tension on the luff of a mainsail, the Cunningham eye was successfully used on the America's Cup defender *Columbia*. Her owner and skipper was Briggs Swift Cunningham, heir to the founding financier of Procter & Gamble, who was better known as a racing driver. An innovative motorsports enthusiast, Cunningham drove at Watkins Glen, and took part at Le Mans in 1952 in the Cunningham C-4R – sometimes claimed to be America's first proper sportscar.

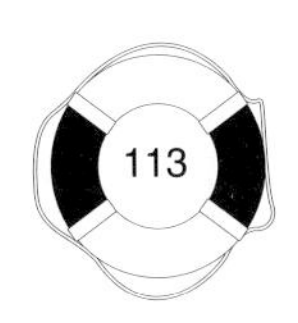

Rex Whinfield's Terylene Sail

Invented in 1941, the revolutionary yarn Terylene extended the range of quality products sailmakers were able to produce. Hitherto, for centuries flax canvas had been the most readily available sailcloth, although it suffered from certain limitations. Sails cut of it were susceptible to discolouring and so needed to be tanned, most commonly with an embrocation of linseed oil, ochre and water. A particularly exotic protective recipe was devised by Claud Worth and published in his book *Yacht Cruising*. In a copper vat he boiled up a mixture of venetian red ochre, raw linseed oil, paraffin and beeswax, and brushed it on to the sails using a paint scrubber; the sails were then hung up for two days in an open shed to drip, then spread outdoors for a fortnight to dry! Untreated flax sails were lighter but extremely porous and some ingenious racing skippers would hoist a crewman aloft to keep the sail moist by pouring buckets of water over the canvas. Whatever was done, flax sails were invariably baggy, became heavy when wet, and when they dried could shrink into erratic shapes. Keeping their appearance and setting better were Egyptian cotton sails. They were originally introduced to Britain in 1851 by the sensational schooner *America*, which winningly had hers close textured and machine woven, and cut to set as flat as possible. But cotton sails had their disadvantages, needing to be carefully stretched when new; and, being more susceptible to mildew and rot, they lasted only some eight or nine years. Worse still, they could rip apart without warning.

For racing yachts the use of artificial fibres solved many of these problems. In 1937 an experimental genoa made of rayon, a cellulose-based yarn derived from processed wood pulp, was created for the America's Cup defender *Ranger*. But a more significant step had been taken back in 1931 when Dr Wallace Carothers, a chemist employed by Du Pont, discovered the first truly synthesised fibre – a polyamide which became known as nylon. It was briefly experimented with as a sailcloth in 1940 but was found to be inappropriate for use on a jib or mainsail as it stretched under severe pressure. Its lightness and stretchability made it best suited for constructing light reaching sails and spinnakers, and by 1946 it was possible for yachtsmen to buy nylon spinnakers made from parachute cloth.

The most significant step forward in synthetic sailcloth development was made in 1941, and in the most improbable of places: an obscure print works in Accrington, Lancashire. In 1923 Rex Whinfield found that, although he had just gained a chemistry degree at Caius College, Cambridge, the only work he could find was unpaid. Luckily, however, it was at the Lincoln's Inn laboratory of Charles Cross, who in 1892 had become the principal inventor of viscose-rayon. Inspired by his proximity to genius, Whinfield became determined to discover a synthetic fibre of his own, but it was not until 1935, when he was running a small laboratory in Accrington for the Calico Printers Association, that Whinfield learnt of the extensive writings of Wallace Carothers. Awash with the funds accumulated from manufacturing wartime explosives, the giant chemicals concern Du Pont had head-hunted Carothers to conduct fundamental research, in the hope it would eventually lead to radical new products. Along the way he discovered nylon. But much to Du Pont's unease, the academic Carothers insisted on publishing his method. Thus Whinfield realised that Carothers's team had decided to concentrate on polyamides and had not much bothered with polyesters, which could be formed by acid and alcohol condensation. With virtually no resources, and aided only by another researcher, James Dickson, Whinfield began exploring polyesters further by experimenting with terephthalic acid – a constituent of fabric dye – which he combined with various forms of alcohol. Success eventually came in the spring of 1941 when Whinfield dipped a glass needle into a thick liquid in a test-tube and

Rex Whinfield's invention of Terylene began to revolutionise sailmaking in 1952, when the elegant 8-metre *Sonda* became the first British yacht with Terylene sails. (Douglas Young)

drew off a thin fibre a few inches long. Effectively he had made the first Terylene fibre by mixing dye with ethylene glycol – commonly used as car anti-freeze!

The new wonder fibre – technically polyethylene terephthalete – was patented by Whinfield in 1942 as 'Terylene'. It was developed for Calico Printers by Imperial Chemical Industries (ICI), who produced their first Terylene yarn in 1946. Extensive tests at the National Chemical Research Laboratory on the banks of the Thames at Teddington showed that Terylene was durable enough to hold its shape and be almost inert in a marine environment. ICI expected that one of the biggest commercial uses for the new product would be for window curtains, but they hoped it could also be used for ropes and sailcloth. A Terylene Council – on which Whinfield served – was established by ICI to market the yarn, but they were limited to making just 50 tons of sample batches per year until the huge works being constructed at Wilton on Teeside came on stream in 1954. In late 1951 Gowen & Co., sailmakers of West Mersea, Essex, were approached by ICI and encouraged to pioneer the first sails made of Terylene. That winter, in conjunction with ICI, Gowen's developed a sailcloth from which they made a genoa. The first commercial Terylene sail ever made in Britain, it was used on a new James McGruer-designed 8-metre cruiser-racer called *Sonda* in the spring of 1952.

Ironically, Carothers's failure to discover Terylene meant that in 1946 Du Pont had to buy Calico Printers' American patent application. They then produced their own pilot samples of the yarn at a plant in Delaware. It was given the trade-name 'Dacron'. The first American sailmaker to develop Dacron was the Marblehead-based Ted Hood, who, reckoning it had potential to provide a tighter weave – and thus far less air friction – than existing sailcloths, purchased looms to weave his own fabric. A year after Dacron became commercially available, in 1954 it had its first racing success when sails made of it helped win the Star world championship. The first ocean racer fully suited with Dacron sails was *Carina*, which won the 1955 Fastnet race.

A gentle and modest man, Whinfield became a great ambassador for the product he had created, and finished his career in 1963 on the board of ICI Fibres. By then Terylene had replaced all forms of sailcloth in every major British racing class, the Firefly being the last to switch to Terylene in 1960. However, despite its numerous qualities, and the ease with which it could be stowed, Terylene was never totally resistant to sunlight deterioration, and that meant boats with Terylene sails had booms with covers and furling jibs with ultra-violet protective edges.

Hood Sailmakers, www.hood-sails.com; Gowen Ocean Sailmakers, www.gosails.com; and McGruer Boats, www.mcgruer-boats.co.uk.

The Yacht Marina

Begun in 1939, the Washington Marina became the world's first yacht marina. President Franklin D. Roosevelt had ordered its construction to encourage the citizens of America's capital to sail for leisure. It was designed by Charles Chaney, a Philadelphia harbour engineer who subsequently popularised the marina idea by writing influential architectural books showing how they could be created. The world's largest artificial yacht harbour is now Marina del Rey on the Los Angeles coast, which has more than 6,000 berths!

Peter Chilvers's Sailboard

In 1958 the first steerable sailboard was launched in Chichester Harbour, from a beach on Hayling Island, by its inventor Peter Chilvers, then aged just twelve! The fully decked, low-tapered hull was exactly 8ft long because that was the size of the only two sheets of plywood the young inventor had to hand. These and all the other components were donated by his mother. With them Peter created twin booms (broomsticks securely tied together to form a wishbone), a mast (formerly a pole used as a roller for a shop blind) and a triangular Terylene sail (cut from a tent groundsheet). The boat had a simple dagger-board, and for steering there was a rudder with a loop on the tiller to enable Chilvers to steer with his foot while sailing the boat standing up. However, the more he sailed the craft at Chichester – sometimes for nearly a mile across to Thorney Island – the more he began to think the rudder was unnecessary because he could steer by tilting the rig. Named the 'Sailboard' by its patron, Mrs Chilvers, the frail craft was used for four summers until its hull deteriorated. Built of domestic-quality plywood, it delaminated beyond repair.

Chilvers never patented or publicised his invention, nor for many years was it commercialised in any form. Thus it was entirely unknown by a Pennsylvanian inventor called Newman Darby, who in 1964 built a bulky oblong board measuring 10 feet by 3 feet to which he attached a kite-shaped sail. An early attempt made by Darby in 1948 to power a small catamaran with a similar hand-held sail had been unsuccessful because of the inefficiency of the rig. Darby's 1964 craft, which coincidentally he also called a 'Sailboard', sailed downwind like a square-rigger but otherwise steered awkwardly. Nevertheless, Darby reckoned it could be a winner. In 1964 he began producing and marketing the boards, and in an article in the August 1965 edition of *Popular Science Magazine* he enthusiastically promoted the new watersport he called 'Sailboarding'. But Darby's design suffered from rather antiquated sail technology and was constructed too heavily, and despite all his efforts sales were unexpectedly disappointing. By 1966 only 80 had been sold in a completed form and his company Darby Industries ceased building them because they were commercially unviable.

A chance conversation in California between Hoyle Schweitzer, a keen surfer, and a dinghy sailor called Jim Drake was to revolutionise the new sport of sailboarding. How could it be possible, they wondered, to surf when there were no waves? Drake, an aeronautical engineer, was driving to a meeting with the US Air

FAMOUS FIGUREHEADS

VESSEL	FIGUREHEAD
Cutty Sark	Nannie the Witch
Golden Hinde	Gilded deer
Mary Rose	Unicorn
Revenge	Lion
Sovereign of the Seas	King Edgar

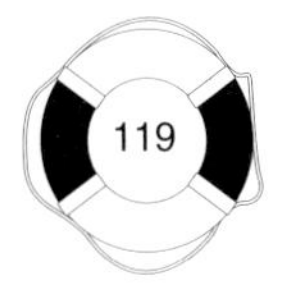

Force when he suddenly realised how it could be done – a sailable surfboard! According to an article on the history of windsurfing by Gregg Dunnett of *Boards* magazine, a 70lb prototype was successfully tested in 1967, and Drake and Schweitzer then patented their 'Windsurfer' in the United States in March 1968, and in Britain in February 1969. Patents were also taken out in other countries, including Germany. Initially even this form of boardsailing was not popular in America and Drake sold some of his patents to Schweitzer, but it was a costly mistake as a tidal wave of enthusiasm for windsurfing then began to sweep through the coastal resorts of Europe. By 1978 some 150,000 sailboards had been sold in Europe, and Schweitzer became increasingly active there issuing writs against makers who were producing them without Windsurfer International licences.

Chilvers himself was increasingly concerned that such litigation was impeding the development of the sport. Indeed, Schweitzer's insistence that only boards approved by Windsurfer International could be used in the forthcoming 1984 Olympic Games threatened to jeopardise windsurfing's very inclusion in the Olympics as a sailing class. Such was the climate of apprehension that in 1981, when a Chilvers-designed board called *Scimitar* achieved a world speed record for a production boat at 21.7 knots at the Weymouth Speed Trials, even he wondered if there might be some legal action. An opportunity to clear the air presented itself when Windsurfer International issued a writ against Tabar Marine, a British company owned by Bic, for importing Dufour sailboards made in France. Cooperating with Bic's defence team, Chilvers revealed publicly for the first time that *he* was the original inventor of the sailboard. If he could establish that his sailboard had been built before Windsurfers' patent was granted, and that it had the same elements, the right to produce such boards would become public property. Evidence given in the witness-box by Mrs Chilvers and former Hayling Island holidaymakers proved crucial in the High Court, which in 1982 overturned Schweitzer's patents on the grounds of 'prior art'.

Yet why was there so much fuss? The concept of sailboats being steered without rudders was nothing new. In 1911 Frederic Fenger had toured the Caribbean in a rudderless sailing canoe called *Yakaboo* which he steered by trimming three sails, and in the 1800s near New York there was a type of boat called a Scooter which was steered with its jib and by moving the crew. Nor was the kite-type sail new. For centuries they had been used by ice-skaters in Baltic countries, and even in the United States, where in the eighteenth century a particularly enthusiastic ice-kiter was none other than Benjamin Franklin.

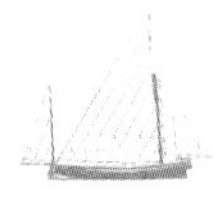

A replica of Peter Chilvers's prototype sailboard is at the Royal Victoria Docks Watersports Centre, London, www.victoria-dock.com; and Newman Darby's windsurfing picture collection is at www.americanhistory.si.edu.

5

Keels and Hulls

William Petty's Catamaran

It is generally assumed that catamarans were invented by the ancient Polynesians and only made their appearance in Europe in recent decades. In fact, totally independently of what went on in the Pacific, some 340 years ago the catamaran was invented in Britain, although the idea was then mysteriously abandoned.

By profession the Hampshire-born Sir William Petty was a military surgeon, first serving with the Royal Navy and then in 1652 becoming physician-general to the army in Ireland. After the Restoration he gained the confidence of Charles II as a cartographer (he fully surveyed Ireland for taxation purposes), but it was as a distinguished statistician and economist that he became best known. It seems that he was pondering on how the Army in Ireland could be supplied by sea when it occurred to him that vessels for such a purpose would need to have long narrow hulls. After doing some simple tank-tests with models, he decided that a radical twin-hulled structure was required in order for such vessels to obtain stability and sail-carrying ability.

He submitted his ideas to Charles II in an ambitious memorandum in early 1662, and having gained royal consent he began building a prototype in Ireland. *The Simon and Jude* was of surprisingly modern appearance. Launched in October 1662, she eventually had two hulls of cylindrical section in a 20ft-long parallel body, with 'heads' or bows each 5ft long. A row of posts was fitted down the centre-line of each hull, and across the hulls (or 'cylinders' as William Petty called them) was a wooden deck with hand-rails above. A 20ft mast was stepped on the forward crossbeam. For such an unconventional vessel she went well but Petty remarked that 'her tails were wanting (being directly flat and barrel-headed behind) the effect thereof was that she drew as much dead water as hindered her and which made a noise like that of a mill tail'. He decided to make alterations to the rig and to fit tails. In a letter to the Royal Society dated 15 November 1662 he remarked: 'Since my last of 29 October we have fitted on those stern pieces so as the whole is now full 30ft. This hath fully answered our intentions for there is now no dead water or noise behind.' During these trials, which Petty describes in detail, it is clear that *The Simon and Jude* sailed superbly well and her crew of four estimated that on a broad reach in a very strong wind she had approached 20 knots! Petty's extraordinary first effort bears a remarkable resemblance to modern catamarans, and in fact won the first race in which a catamaran took part. On 12 January 1663 *The Simon and Jude* decisively won an open sailing race, actually lapping several monohull competitors.

Emboldened by this success, in July 1663 Petty designed a larger version. Built in Dublin, *Invention II* had two 30ft hulls which were 4ft wide and 8ft deep. Curiously, Petty now abandoned the cylindrical-hull configuration which had worked so well, and opted instead for hulls that looked as if they were the separated halves of a conventional vessel. Nevertheless, *Invention II* was also quick. During her second voyage from Holyhead to Dublin she won a £50 wager by easily outsailing the fastest packet boat the Royal Navy had in those waters. Indeed, this was the first recorded race off the British coast in which a yacht had participated. Samuel Pepys even favourably recorded news of the catamaran yacht's victory in his famous diary.

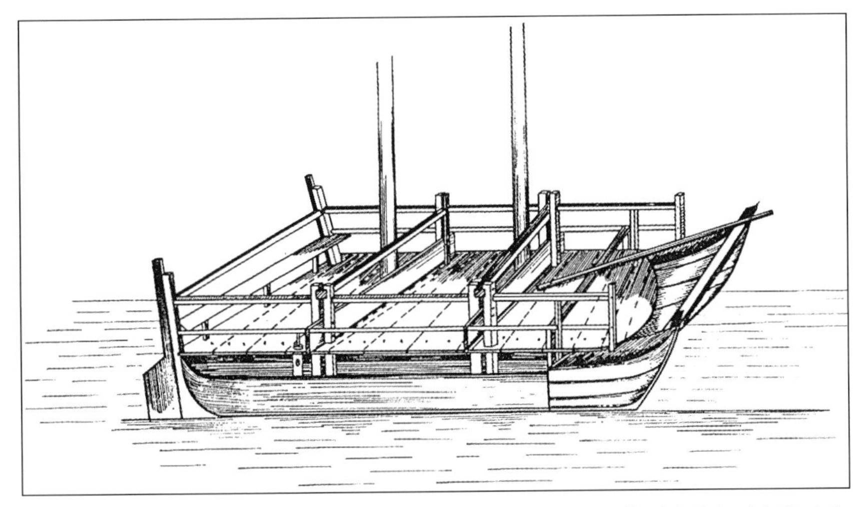

This experimental catamaran created by William Petty in 1662 reputedly did 20 knots! (Corlett, *Transactions of the Royal Institution of Naval Architects*, October 1969)

Petty must have reckoned his revolutionary twin-hulled invention was on a roll when the *Experiment*, an even larger catamaran, was launched in the presence of Charles II, John Evelyn and Pepys at Rotherhithe in December 1664. Setting sail for Oporto in April 1665, the *Experiment* was accompanied to Lands End by frigates which she outsailed. It was a fine achievement given that 33 of her 50-strong crew had been 'pressed' by the Navy. But her luck changed on the return trip, when she tragically foundered with all hands in the Bay of Biscay during a ferocious storm. Quite why she was lost is unclear. Certainly by then Petty must have realised the major disadvantage of any catamaran was that in a strong wind or a squall it was likely to lift its weather hull out the water, and beyond a certain point it would become unstable and capsize. Even so, by reaching Portugal the *Experiment* had proved she was just as seaworthy as more conventional monohulled ships – some seventy of which were reported to have perished in that same ferocious storm.

Uncertain why the *Experiment* had sunk, Petty seemed to lose the plot. Rather than going back to the smaller catamaran designs which had done so well, he unwisely pushed ahead with an even larger innovative vessel. But *St Michael the Archangel*, which he designed in 1684, was a disaster. Of 128 tons burden but only 20 feet in total beam, she was far too narrow and full, with dismal sail-carrying ability. Although she had a 'W'-shaped single hull, and was thus not a proper catamaran, she was described at the time as having a 'double keel', which meant her failure quite unjustly established the myth that Petty's fourth catamaran was considerably less successful than his first. His detractors then argued that throughout the course of the development programme his catamaran designs had somehow progressively got worse.

Sir William Petty, the mapmaker who became Europe's first catamaran designer. (Petty, *Hibarniae Delineatio*, 1685)

One particularly influential critic was the ingenious navy commissioner Peter Pett, who was the master shipwright at Chatham. Perhaps jealous of Petty's accomplishments, he tried to discredit the catamaran concept in a letter to Samuel Pepys: 'Petty's boat is the most dangerous thing in the world. If it should be practised it would endanger loss of our command of the seas and our trade. If the Turks and others shall get the use of them, and by bearing more sail will go faster than any other ships, our merchants will be at the mercy of their enemies.' Peter Pett was officially reprimanded in 1667 for failing to prevent a raid by the Dutch fleet on the Medway, an event that demonstrated how disastrous it would be if the Netherlands got the idea of building shallow-draft fighting catamarans. Charles II himself now noted that he should be sorry if the catamaran invention succeeded, 'for then the Hollanders would have as much advantage of us as we have now of them'. The loss of royal endorsement effectively meant the end for Petty's catamaran. Having twice refused a peerage, Petty was not well-known to people of money and influence, and so the catamaran idea lacked support from a wider circle and petered out.

It was not until some 270 years after Petty's death in 1687 that sailing catamarans were again seen in Britain. This time the inspiration for such boats really did come from Hawaii, where, just after the Second World War, an aircraft engineer called Woodbridge Brown modified a traditional Polynesian design of catamaran to build a 40ft craft called *Manu Kai* using thin marine ply. Crucially, and entirely unlike Petty's design, the outermost sides of her hulls lay almost dead straight, while the inner sides had a marked curvature. With the crucial lateral resistance provided by the straight lines of the hull, the effect was a large stable catamaran which reportedly could reach speeds of 20 knots. This improved Polynesian-type of yacht design was introduced to Britain in the 1950s by a boatbuilder called James Wharram, but it was not until 1973, when Austrian-born Wolfgang Hausnet successfully completed the first circumnavigation of the world via Panama in a plywood catamaran, that William Petty's faith in the seaworthiness of sailing multihulls was belatedly vindicated.

Yacht plans of traditional Polynesian-derived catamarans are at jameswharramdesigns.co.uk; a full-scale replica of *The Simon and Jude* is at the Conservatoire International de Plaisance, Bordeaux; and Petty's model of *Invention II* is at the Royal Society, London.

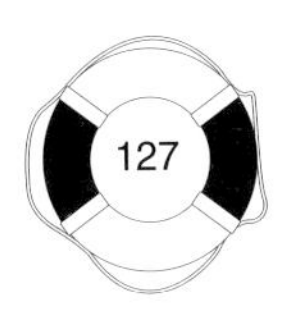

John Schank's Dagger-board

In 1776 exceptional mechanical ingenuity was used by a Royal Navy lieutenant in order to attack the forces of General Benedict Arnold, one of George Washington's most trusted officers in the War of American Independence. Lieutenant John Schank arranged for a brig called *Inflexible*, and a few other fighting vessels, to be dismantled and moved overland by means of capstans and rollers from the St Lawrence River to the rebel-held Lake Champlain in Vermont. That October, with the vessels all reassembled and with Schank himself in command of the flagship *Inflexible*, they soundly defeated Arnold's fleet.

It was while stationed in Boston, Massachusetts, that Schank thought of a method of manoeuvring vessels in shallow water using sliding keels. In 1771, with the permission of the senior officer there, Earl Percy (later the Duke of Northumberland), he built a small prototype craft. Her lines were similar to those of local boats, but Schank had secretly installed a dagger-board that extended for two-thirds of the length of her keel. He reckoned that the lowering of such a wooden plank through a slit cut in the keelson would prevent leeway when the vessel was close-hauled. That had been the purpose of the leeboards fitted to the sides of Dutch boats for centuries, and indeed for thousands of years on Chinese junks. Schank's invention differed in that he was the first person in the western world to advocate the use of a keel that would slide through the centre of the boat and could be raised when not in use.

On his return to England Schank was promoted to post-Captain and was appointed Dockyard Commissioner at Deptford. In that capacity he neatly devised a scheme for cutting grooves in gundecks to enable cannon to be trundled rapidly across to whichever side of the ship they were needed, and he also developed a system of inside shutters to prevent gunports leaking. For the purposes of evaluation, in 1775 the prototype sliding keel craft was freighted to England, and

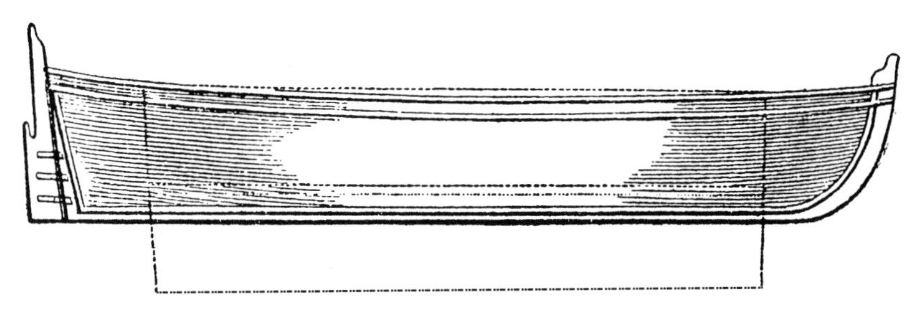

The first dagger-board Schank fitted in 1771 cut through most of a Boston longboat's keel. (Charnock, *An History of Marine Architecture*, 1802)

it thus became the first such boat in British waters. Schank improved its versatility and steerability by replacing the single long sliding keel with three dagger-boards, and even wrote a short treatise on the subject. By 1789 he had bamboozled the Admiralty into putting his sliding keel idea to the test on the Thames by installing three adjustable dagger-boards in a 13-ton flat-bottomed sloop. This duly outsailed a similar boat that had leeboards. Indeed, it so outdistanced the other that in 1791 a bigger sliding-keel vessel was ordered. This was the 68ft revenue cutter *Trial*, which Schank fitted with watertight bulkheads. She proved so successful that by 1797 the Admiralty had been fitting sliding keels in various brigs.

A particularly notable warship of that type was the 60-ton *Lady Nelson*. Built under Schank's supervision at Deptford, the capabilities of her adjustable keels won her the task of charting in detail the hazardous south-east coast of Australia. This meant a two-year expedition, during which she is said to have become the first British vessel to have sailed around Tasmania. However, it was not all plain sailing for the *Lady Nelson*. Like the other sliding-keel naval vessels, she suffered serious leakages of sea-water through the dagger-board casings and the boards were apt to get stuck. The talented Schank became a vice-admiral in 1810. But beyond his brilliant watertight bulkhead idea, even he could not find a way to solve those

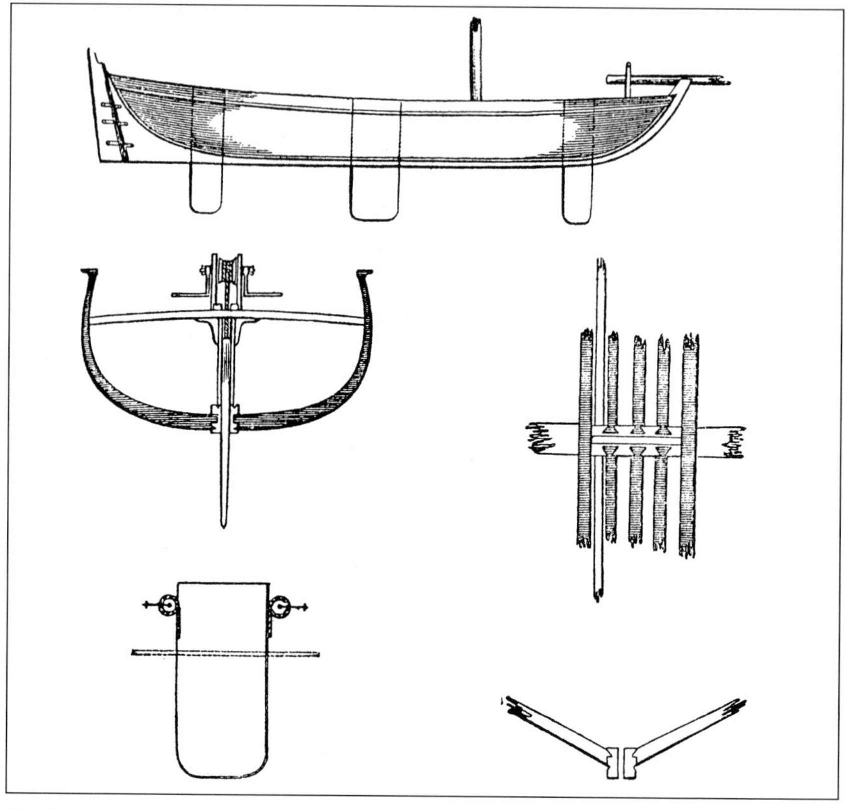

This demonstration sailing vessel Schank built at Deptford in 1778 had three dagger-boards. In the centre right diagram, the long dark lines on either side of the dagger-board case are examples of Schank's newly introduced watertight bulkhead. (Clark, *The History of Yachting*, 1904)

potentially dangerous problems; and soon the Royal Navy – and indeed the British mercantile marine – became so disenchanted with the sliding keels that they ceased to fit them.

The solution, belatedly discovered in 1809 by Captain Molyneaux Shuldham RN, was to use a centrally positioned triangular board that pivoted on a bolt through the fore part of the boat's keelson. Shuldham proposed that what he called his 'revolving keel' should be housed in a

Admiral John Schank, inventor of the dagger-board. (*European Magazine*, 1805)

watertight wooden case, ribbed with copper or zinc for added strength; this would decrease friction and avoid the issue of the keel getting jammed. At the time Shuldham was a prisoner-of-war, being held by the French at Verdun, but somehow the original model he made of this cheesecutter-like contraption in 1809 found its way to London, where it was exhibited at the Adelaide Gallery. There it may have come to the notice of the three Swain brothers from New Jersey who in 1811 patented it in the United States as the 'Centre-board'.

The first commercial centreboarders of any size were sloops sailing on the Hudson River, but from there the design spread along the vast stretches of shallow water on America's East Coast and soon appeared on 70–80ft working schooners. Shunned though centreboarders were by Britain's professional sailors, the device became a standard feature on around 80 per cent of America's enormous coasting fleets, on all types of craft including scows, skipjacks, bugeyes, catboats and even the Chesapeake oyster sharpies. And in due course some centreboarders, notably the New York 'sandbaggers', evolved into speedy racing machines.

The first yachtsman known for certain to have used a sliding keel was Commodore Thomas Taylor of the Cumberland Sailing Society, who in 1794 had a boat with no fewer than five of them built for racing on the Thames. The *Cumberland IV* competed with some success, but concerns about the leaking cases persisted and few other British racers were willing to adapt their conventional keels. Such scepticism seemed justified in July 1876 by a dramatic accident involving the *Mohawk*, a 141ft schooner whose precarious 6ft draft was made safe by a gigantic 31ft centreboard. Anchored off Staten Island during the preliminaries of the America's Cup, she was just setting sail for an afternoon's leisure cruise when she was hit by a sudden squall. Her centreboard had not been lowered and *Mohawk* capsized and sank. Tragically her owner, the vice-commodore of the New York Yacht Club, and all his guests were lost, some trapped below by shifting furniture.

Acceptance of the centreboard in Britain was largely restricted to dinghy sailing, but developments emerged in the 1900s from the world of sailing canoes by means of clever devices such as the fan-like Radix folding centreplate. So widespread was the use of the centreboard in America that people automatically assumed it was an American invention! In Britain the Scottish-born Schank has never been honoured for his pioneering idea, but in Australia there is a lasting tribute to him – a distinctive landmark on the shallow south-east coast named Mount Schank.

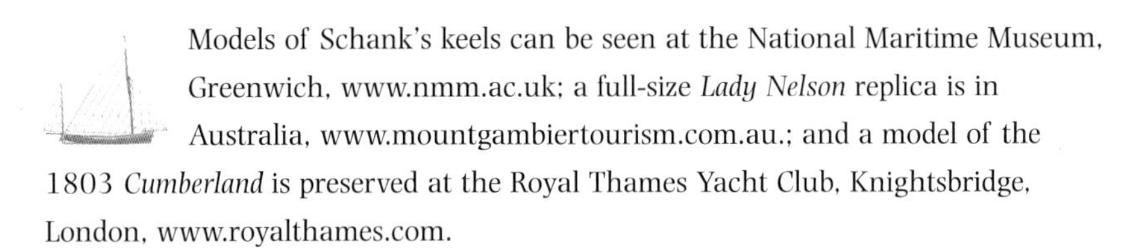

Models of Schank's keels can be seen at the National Maritime Museum, Greenwich, www.nmm.ac.uk; a full-size *Lady Nelson* replica is in Australia, www.mountgambiertourism.com.au.; and a model of the 1803 *Cumberland* is preserved at the Royal Thames Yacht Club, Knightsbridge, London, www.royalthames.com.

Patrick Miller's Trimaran

Many centuries ago Polynesian islanders reckoned the stability of their canoes could be enhanced by attaching crossbeams with floats,

The Land Yacht

Originally developed by the ancient Chinese, and then reinvented by the Dutch in the seventeenth century, the land yacht was perfected by Captain Molyneaux Shuldham RN (better known as the inventor of the swivelling mast). In 1809, banned by French guards from boating on the River Meuse at Verdun where he was held captive, Shuldham amused his fellow prisoners-of-war by building two sophisticated land yachts – a schooner and a sloop – and sailing those instead!

creating what they called a Latakoïs. The term 'Trimaran' was only coined in the 1940s, by a naturalised American called Victor Tchetchet. In the meantime, in 1786 to be precise, a Scotsman called Patrick Miller became the first European to design a three-hulled boat.

By that year Patrick Miller had accumulated a substantial fortune as a merchant banker in Edinburgh, which enabled him partially to retire from business, but as a man of great energy he intended to devote himself to virtuous accomplishments. Initially he was concerned with agricultural innovations. At his Dalswinton estate in Dumfriesshire, which he purchased in 1785, he was reputedly influential, introducing Scotland's first drill plough and threshing mill, fiorin grass from Ireland, and even the first turnip seeds – which were sent to him by King Gustav of Sweden (hence the name 'Swede' for a type of turnip). Moreover, because he had spent some of his earlier years at sea, he was keen to find a means of enhancing Britain's naval strength. As a principal shareholder in the famous Carron Iron Company, Miller had already been influential in building the fearsome short-barrelled and light Carronade. Keen to develop a reliable means to facilitate the effective tactic of 'breaking the line' – newly established by Lord Rodney off Cape St Vincent in 1790 – Miller devised a form of

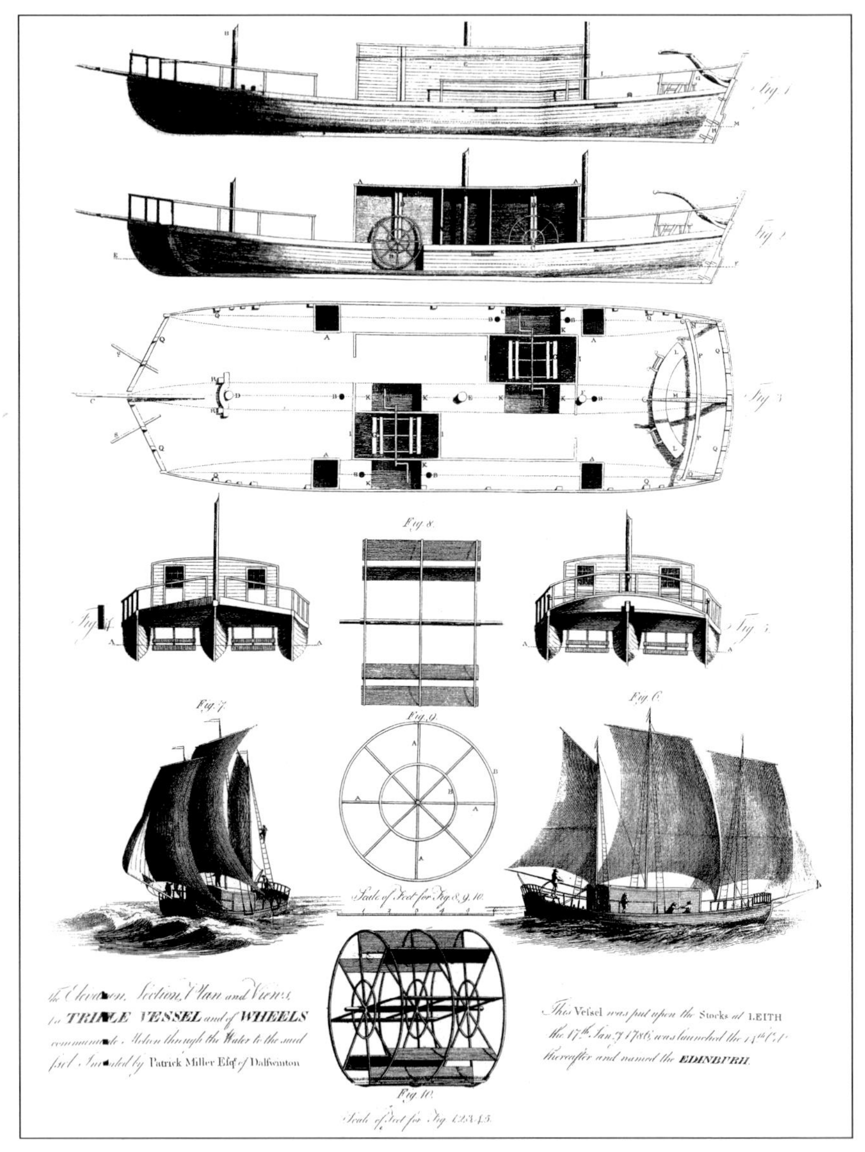

Edinburgh, the trimaran Miller built in 1786, was intended to be propelled with paddle-wheels. (Miller, *The Elevation, Section, Plan and Views of a Triple Vessel*, 1787)

paddle propulsion which would enable the Royal Navy to strike at becalmed enemy ships without needing to rely on sail.

Optimistically assuming the paddle-wheels could be turned by a five-bar capstan powered by crewmen on deck, Miller calculated that what was needed for the demonstration trials was a vessel broad enough to accommodate the capstan operators and wide paddles, yet also hydrodynamically efficient. To create what was effectively a giant pedalo he opted to have three hulls. Construction of the sensational so-called 'triple ship' *Edinburgh* commenced at Leith in January 1786 and she was launched that October. Miller lavished substantial funds on her development and even wrote a prospectus entitled *The Elevation, Section, Plan and Views of a Triple Vessel with Wheels* which included technical drawings by the society portraitist Alexander Nasmyth.

On 2 June 1787, within an hour of the initial experimental trial beginning on the Firth of Forth, Miller was becoming despondent. Despite the calm sea state progress had been slow and the capstan-turners were already nearly exhausted. Just then James Taylor, a tutor accompanying Miller's two sons on board, made a suggestion: 'Perhaps try the power of steam, Sir.' Inspired, Miller decided to employ as a consultant a struggling inventor called William Symington who had been experimenting with a steam engine designed for a road vehicle. However, as Symington's engine required two paddle-wheels to be run in series, the trimaran *Edinburgh* was discarded for a new multihull made of tinned iron. But that new 60ft steam-powered catamaran yacht, tried on Miller's private lake at Dalswinton in October 1788, soon failed because of Symington's decision to use an unreliably complicated chain-drive mechanism. Then in December 1789 a replacement steam catamaran's paddle-wheels disintegrated. Furious at Symington's apparent incompetence, and having wasted a fortune on his various experiments (estimated in 1885 to have been £30,000), Miller was inclined to let the question of steam propulsion drop. Not until 1803, with a different patron and the monohull *Charlotte Dundas*, was Symington convincingly able to demonstrate steam's reliability.

Curiously, Patrick Miller had not sought to make money from his mechanical trimaran. Although it was intended to be a weapon of war, he soon saw it could have more peaceful purposes, as a shallow-draft lifeboat or a cargo boat. Instead of applying for a patent, in 1786 he distributed the prospectus to all the sovereign heads of state in Europe with the intention that the trimaran should be used for the good of the world. Miller returned to professional life in 1790, becoming deputy-governor of the Bank of Scotland. Although he had abandoned the idea of steam navigation, he was still interested in the improvement of naval architecture. In 1796, perhaps because of the pressures of war, he quietly applied for a patent for a device to propel multihull and shallow-draft vessels, powered not by steam but by horses! But such was the growing preoccupation with steam it was never fully developed and was largely ignored.

Instead, the three-hulled boat would be reinvented many years later by a Bohemian artist better known for his pastel drawings of ladies who modelled for romantic book covers and top-shelf gentlemen's magazines. Life had at first been far from artistic for Victor Tchetchet, an Olympic competitor who in 1908 had built a catamaran from two kayak hulls and entered it in the Spring Race of the Imperial Yacht Club at Kiev. The boat won easily, but was so radical she was subsequently disqualified and permanently barred. In the First World War Tchetchet joined Imperial Russia's air service as a pilot, and the subsequent revolution caused him to flee to New York's Long Island, where he made a living as an illustrator. Such was his burgeoning fascination with multihulls that by 1943 he had invented the western world's first triple-hulled simple *sailboat*. Launched in the summer of 1945, this 24-footer, which he called a

'Trimaran', looked rather like a canoe with outriggers and had a conventional sloop rig. The next year Tchetchet proved she was faster than more conventional boats of similar size by winning second prize in the Marblehead Race week.

Patrick Miller. (Science Museum/Science and Society Picture Library)

There was another reason why 1946 was the year in which the yachting trimaran really came of age. That September a 43ft junk-rigged trimaran called *Ananda* set out from Casablanca to cross the Atlantic. André Sadrin, her French owner, found the ketch's motion to be uncomfortably quick; worse, she became awkward to steer when her floats leaked, and so low was her freeboard that her decks were frequently awash. But she was sturdily built of oak and mahogany, withstood everything the Atlantic could throw at her, and in early November successfully reached Martinique. Some 24 years later the high-tech trimaran *Paul Ricard*, skipped by Eric Tabarly, captured the speed record for sailing across the Atlantic in a time of just 10 days 5 hours. This historic success stimulated others to make record-breaking attempts in multihulls.

Ananda's design, at least in terms of hull shape, was strikingly similar to that of the original trimaran *Edinburgh*, proving that Miller had been essentially right. Indeed it is ironic that multihulls such as trimarans, which Miller had regarded as scarcely more than experimental platforms upon which to develop more efficient means of propulsion than conventional sailpower, became some of the fastest sailing yachts of all. If only Patrick Miller had not got involved with steam!

Further details on multihull designers and the history of trimarans at www.multihull-maven.com

Edward Bentall's Fin Keel

Improbable though it might seem, the person who devised the fin keel, a feature common to most yachts now, was a Victorian plough-maker! Edward Bentall was an audacious rural engineer who, on inheriting the family agricultural machinery business, wondered how its mainstay, the simple 'Goldhanger' plough invented by his father, might be improved. Aware that friction greatly restricted the plough's efficiency, he overcame the problem by suspending the plough blade beneath a tricycle frame. As it caused less ground friction, this frame was able to carry more than one blade and thus could cut several furrows simultaneously. Awarded a gold medal at the 1851 Great Exhibition, Edward's revolutionary patent 'Broad-share' cultivator sold throughout the world and became such a commercial success that it required the huge expansion of the Bentall factory and the dock basin on the River Blackwater, near the Essex coast. This thriving plough business provided the financial wherewithal for Edward Bentall, a keen yachtsman who had already designed a few centreboard boats with marked success, to fund his boundless enthusiasm for radical yacht design.

In the 1870s it was still conventional for yachts to have 'cod-head and mackerel-stern' hulls, and straight keels which were almost as long as the vessel's waterline length. In 1874 a pioneering

The Ice Yacht

The largest ice yacht ever was the 68ft *Icicle*, which was built for Commodore John E. Roosevelt (FDR's uncle) in 1870 to sail on the frozen Hudson River. A gaff-rigged sloop with 1,070sq ft of canvas, *Icicle* was so phenomenally fast that in 1871 she raced against the Chicago Express steam train, and won!

hydrodynamicist called William Froude, having carried out tests for the Admiralty on model warships in a giant test-tank in Torquay, revealed to the scientific world he had discovered the principal cause of resistance to any vessel: hull surface friction. This momentous finding would greatly influence the underwater shape of future ship designs. But even for most professional yacht architects the theory seemed impossible to utilise in boat design. A rare example of those who dared to consider applying it to the keel shape of racing yachts was the eminent Clydeside yacht designer George L. Watson. Even he was cautious, producing only a 5-tonner called *Clotilde*, whose underwater shape was merely rounded off. It was then that a quantum leap in yacht design was made by Edward Bentall, the enthusiastic amateur.

Drawing on his practical experience of enhancing ploughs by reducing friction, Bentall wondered if a yacht's wetted (frictional) surface could be reduced by making the forward part of the keel curved. Accordingly he set about devising a yacht which would have a great waterline length to develop speed but would carry her ballast low to attain stability. The result in 1875 was the sensationally radical 99ft yawl *Jullanar*, which Bentall built in the plough factory's Heybridge Basin. Even her metal fittings were made there. *Jullanar* immediately startled the yachting world by demonstrating that the drastic cutting away of her forefoot – hitherto assumed to be sacrosanct – did not impair her performance to windward, which

Plough-maker Edward Bentall's radical 1875 yawl *Jullanar*. (Chatterton, *Sailing Ships*, 1909)

was excellent. Indeed, she proved to be phenomenally fast on all points of sailing, as well as being a sturdy sea-boat. Bentall never raced her himself, but in other hands she reputedly won more races than any other yacht during her racing life.

Emboldened by *Jullanar*'s success, Bentall tried to reduce surface friction still further in a boat with a slender beam and a fin keel. The result, which Bentall built at nearby Wivenhoe in 1880, was an extraordinarily advanced 10-tonner called *Evolution*, which at 51ft long had a beam of only 6ft 5in! Ingeniously the 5-ton lead ballast fin keel was fastened to an inner metal keel that ran the length of the yacht.

Jullanar had been built because, unlike most designers, Bentall had both the audacity and the financial means to be creative and to experiment, and his innate understanding of sailing made up for his basic grasp of naval architecture. But with *Evolution* it all went wrong. As soon as the miracle fin-keeler took to the water it was clear Bentall had seriously miscalculated. *Evolution*'s stability was atrocious, and the fin keel itself seemed to be the cause of her tendency to heel alarmingly in even the slightest breeze. Various improvements were tried but to no effect. Baffled as to how he had so grossly over-reached himself, Bentall immediately ceased all work on the fin keel.

Indeed, virtually no one made a determined attempt to produce a viable fin keel until 1891. That autumn, at a Newport (Rhode Island) boatyard, the society yacht designer Captain Nathaniel Herreshoff launched a bulb-keeler called *Dilemma*. In contrast to Bentall, Herreshoff had the technical knowledge needed to make the crucial calculations for a successful fin keel boat. Ideally, he reckoned, it required a waterline length equal to 3.5 beams (*Evolution*'s had been 8 beams). It also needed a free-standing rudder, and to give stability the midship section of the hull had to be almost round. The result was the scow-shaped 38ft *Dilemma*, whose fin keel was a steel plate, at the foot of which was a cigar-shaped bulb of lead. *Dilemma* never raced, but word soon spread that she was exceptionally fast in sheltered waters, and with her spade rudder, set back several feet from the plate, she was easy to steer.

By the 1892 season the most successful half-rater on the Solent was *Wee-winn*, a Herreshoff fin-and-bulb keeler, which won 20 first and 1 second prizes out of twenty-one starts. Soon she was widely

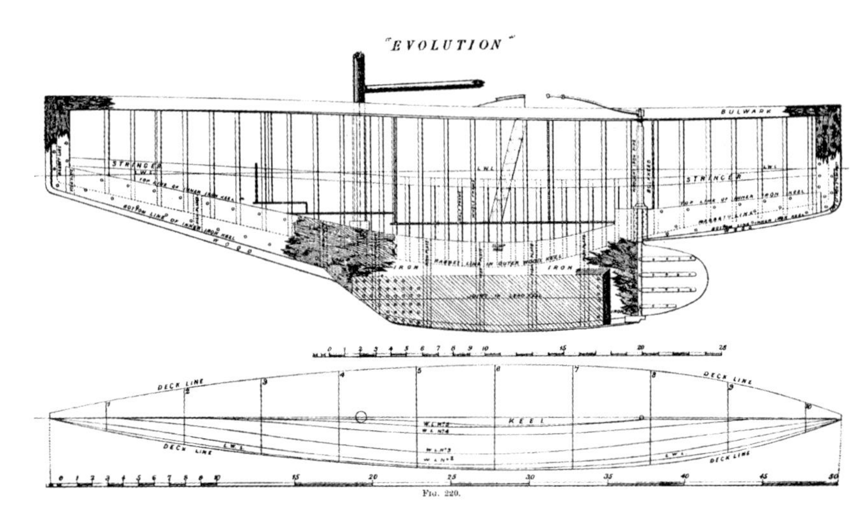

The revolutionary fin keel yacht *Evolution* designed by Bentall in 1880. (Dixon Kemp, *A Manual of Yacht and Boat Sailing*, 1913)

copied. After a campaign against such apparently unseaworthy craft organised by R.E. Froude, a relative of the influential hydrodynamicist, so-called 'skimming dishes' were banned in 1896. And, with that, fin-and-bulb keels effectively disappeared from popular use for the next sixty years.

Herreshoff was careful not to describe himself as the fin keel's inventor but it must have been disturbing for Edward Bentall to learn that British owners had approached the American concern with orders for new yachts. And Bentall never got over his bitter disappointment at *Evolution*'s failure. Indeed, she never sailed again, and not long before his death in 1898 Bentall ordered the historic fin-keeler to be broken up.

A model of *Jullanar*, Edward Bentall's first experimental yacht, is at the Science Museum, London SW7, www.sciencemuseum.org.

Robin Balfour's Twin Keel

Until the advent of the twin keel the only other known forms of side keel were bilge keels. In the mid-nineteenth century it was not uncommon for ships to have so-called 'docking keels' which projected downwards from the hull at the turn of the bilge, in order to support the weight of a large vessel when she was lying on blocks in dry dock. Bilge keels fitted to reduce rolling were sometimes called drift keels. But at that time the only small craft with bilge keels were working boats such as fishing craft, which had them in the form of timbers fastened under their bilges for the purposes of keeping the boat upright when on shore.

The first yacht in Europe known to have had pronounced bilge keels was the Irish ketch *Iris*. Built in Dublin in 1894 for a drinking syndicate, the 60-footer's purpose was to enable them to make a leisurely exploration of all the shallow coves and inlets on the coast between Dundalk and Waterford. To enable *Iris* to settle level on the foreshore (and thus facilitate civilised wining and dining activities) she had a draft of just 3ft 6in, achieved by having two stout bilge keels in addition to the main central lead ballast keel. She seemed to be moderately successful, being a handy craft with an unexpectedly

The Concrete Boat

In 1849 a French horticulturist called Joseph Louis Lambot built the first concrete boat (a rowing dinghy which is now in the Brignoles Museum, Provence). The 'Ferciment' process which he patented in 1855 involved plastering a sand and cement mortar over a framework of iron bars and mesh. *Helsal*, a 72ft Australian ferro-cement yacht completed the 1973 Sydney–Hobart race in the fastest time and even set a course record.

easy motion in bad weather, and able to fetch to windward much better than her owners and some of the local wiseacres had expected. But leading naval architects at the time, such as the famous yacht designer Dixon Kemp, doubted her sailing qualities, believing that all such multiple keels needlessly created a large inefficient area of surface friction.

The unlikely discoverer of a means of gaining the shallow draft advantages of such appendages without the performance limitations was the Honourable Robin Balfour (later Lord Riverdale), who at the time was impoverished and working in a Sheffield steelworks. He had long been keen on sailing, and in 1922, while designing a new boat, he had a blinding flash of inspiration. Why not have twin keels? For that matter, why not twin rudders, too? Balfour tried out his ideas with a towed model in a local lake. He knew there had been bilge keel boats before, but nobody apparently had purpose-designed a twin-keeled yacht without some form of central keel. This concept was revolutionary, as were the twin rudders, the idea being that when heeled the leeward keel and rudder would create more lift.

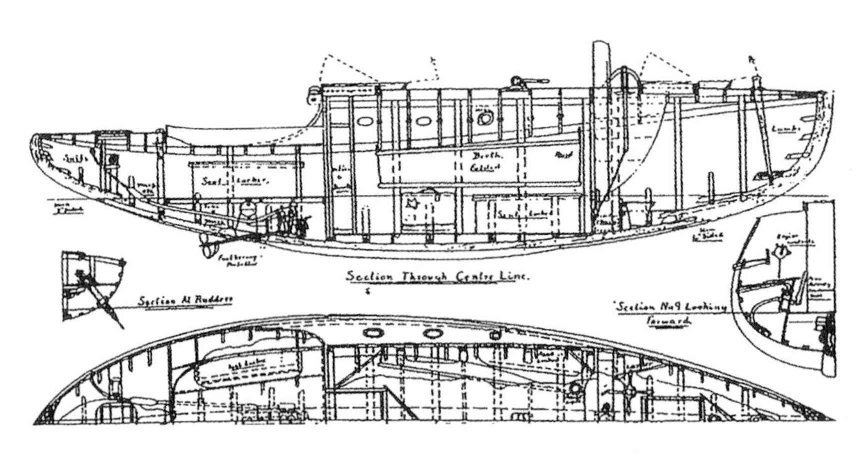

Bluebird, Robin Balfour's innovative prototype. (Riverdale, *Twin Keel Yachts*, July 1968)

Marine Plywood

Immanuel Nobel (the father of Alfred Nobel, the dynamite-inventing benefactor of the Nobel Prize) devised a machine for peeling fine veneers from timber in the nineteenth century. No known glue could hold together underwater, although in 1912 Dr Leo Baekeland, the inventor of Bakelite, suggested that a plastic-type glue might be developed as wood adhesive. In 1934 Dr James Nevins invented a waterproof adhesive suitable for plywood, and reliably waterproof plywood boards were first produced in 1940.

Launched in 1924 on a canal at Tinsley, near where Balfour worked, the 25ft *Bluebird* bristled with innovations. Her well-rounded hull had virtually no keel and was of monocoque construction, like that of an aircraft fuselage. Balfour had designed the unique horizontal arrangement of her winches, and even the vertical stowage of her grappling hook anchor on the foredeck. *Bluebird*'s construction had dangerously emptied his pockets, but the acceptance of a series of articles for *Yachting World* helped transform his financial situation and for the next ten years he cruised thousand of miles in *Bluebird*, circumnavigating Ireland and the British Isles.

Determined to prove that the *Bluebird* twin-keel concept could be applied to larger yachts, in 1939 he built a 39ft version, which he called *Bluebird of Thorne*, and then in 1963 an ocean-going 50-footer of the same name in which he sailed to New Zealand. By then, as Lord Riverdale, he had become Commodore of the Royal Cruising Club. Curiously, twin keels never became popular in America. But in Britain they really took off, even appearing on new boats produced by leading yacht designers such as Robert Clark, Maurice Griffiths, Percy Blandford and Robert Tucker, who would no doubt all have read Balfour's influential earlier articles in *Yachting World*.

Uffa Fox's Planing Dinghy

Avenger became the world's most famous racing dinghy in 1928, winning a legendary 52 firsts, 2 seconds and 3 thirds out of 57 starts. Her skipper, the boatbuilder Uffa Fox, had been born on the Isle of Wight in 1898, the son of a skilled carpenter who had worked on the construction of Osborne House for Queen Victoria. At Cowes Uffa was apprenticed to S.E. Saunders, a boatyard specialising in high performance craft, which even then was developing powerboats capable of moving at over 50 knots, such as *Maple Leaf IV*. Uffa acquired further practical knowledge of aeronautics and of water's limitations on speed during the First World War while serving with the Royal Naval Air Service working on seaplanes and flying boats.

Naval architects had long realised that the factor of water friction upon a hull was so restrictive that, at a speed of about 5 knots, as a boat accelerated, a moment would come when the resistance curve would begin to rise almost vertically. In 1877 a British designer called John Thornycroft had patented a single-stepped skimming hydroplane which would enable boats to break through that barrier by planing on the surface, but this was assumed to be applicable only to fast motorboats. The next step forward came in 1898 when the Seawanhaka Cup was won off New York's Long Island by *Dominion*, an innovative 37ft sailing yacht devised by the American designer Herrick Duggan. *Dominion* was essentially a scow whose underside was so arched she virtually had two hulls, and thus some people wondered if she might have managed to plane by means of creating a cushion of air between her hulls. But her speed was never recorded, and as she was banned from competition (for being a catamaran) she was never developed further. And, as yet, there was apparently still no plane-achieving monohulled sailing boat.

Aged only twenty-one, Uffa Fox established his own boatbuilding business at Cowes, improvising his premises from a disused 'floating bridge' ferry. The central part was roofed over to provide a workshop, while the prow formed a ramp to the shore at one end and the other end became a slipway for launching boats into the River Medina. The former passenger accommodation was converted to a drawing office and living space. He built several types of dinghy there and in the process became intrigued with the new national class created in 1923 by amalgamating the designs of three types of local 14-footer. As the first ever British 'National' dinghy class, a few restrictions were imposed on the Fourteens to enable the boats to be inexpensively moved by rail to race meetings in distant parts of the country. A 14ft hull was the largest that could fit in a standard goods wagon, and reputedly could be sent anywhere in England for 2*s* 6*d*. But the other governing rules for these otherwise traditionally shaped dinghies were fairly vague, so Fox reckoned there was scope for interpretation. Thus in 1925 he designed and built his own National Fourteen. *Ariel* had a conventional 'U'-shaped bow, but incorporated several new ideas of Fox's, including less freeboard than usual, a simplified gunter rig and an overlapping jib. An occasional winner, her successes led him in 1926 to build *Radiant*, an experimental Fourteen with slightly V-ed fore sections. Encouragingly she came second in the first ever prestigious Prince of Wales Cup in 1927.

For the 1928 season Uffa went a stage further, producing a design far ahead of its time. *Avenger* had a fine bow, with prominent V-sections steadily developing into a flat floor. Her greatest depth was at a third of her length from the bow. The V-shaped section of the hull would, Fox hoped, enable *Avenger* to ride over her own bow wave and thus greatly increase her speed. If, that was, Uffa was agile enough to keep her stable and prevent a violent sheer to one side or the other while she developed sufficient horsepower. It

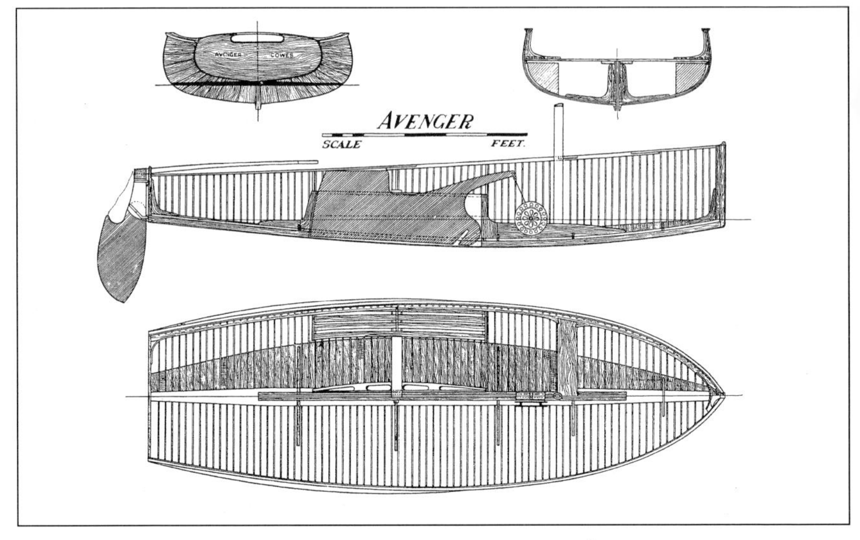

Avenger, the revolutionary planing dinghy Uffa Fox designed in 1928. (Uffa Fox)

worked. The bursts of speed from the planing effect meant that, on the reaching and downwind sections of the races that year *Avenger*'s performance was unsurpassable. An additional advantage she enjoyed was Uffa himself, who was an exceptionally talented helmsman.

He was also foolhardy. That summer Uffa made a snap decision to race at the Le Havre Regatta. Loading *Avenger* with three hundredweight of equipment and a crew of two he sailed across to France. Having won her races there against the best French 14-footers, *Avenger* sailed back to Cowes despite a strong breeze – a hazardous 37-hour voyage in an unlit open boat. On another occasion Uffa challenged another boatbuilder, Frank Morgan-Giles, to race from Cowes to Teignmouth and back for a bar of chocolate! In 1928, the year that *Avenger* won 52 races and finally the Prince of Wales Cup, the National 14 became the International 14. The class duly became the elite of the dinghy world and Fox, with his revolutionary design and sparkling performances on the water, dominated the racing scene until 1939. A substantial order book was soon built up for the famous honey-coloured lightweight dinghy shells being made at his bustling ferry-boat yard. Indeed, after *Avenger* Uffa never advertised. Such had been his headline-grabbing exploits that he was never short of publicity. It established him as the first celebrity dinghy-designer.

During the Second World War he conceived the idea of the airborne lifeboat, a vessel to be slung underneath the fuselage of an aircraft and then dropped to survivors of ditched planes. Constructed with lines that accorded with the shape of the aircraft, the airborne lifeboats contained survival gear, a mast, sails, oars and even an engine!

Avenger was the first true planing dinghy with a good windward performance, and after her success Fox applied the planing concept to various other classes. In 1933, with a V-hulled sailing canoe, he won two national championships in America. In Britain the National Twelve, a class established in 1935 as a less expensive version of the Fourteen, and open to all forms of design, came to be dominated by his planing 'Uffa King' design. Of similar specifications was the Firefly. Selected as a class for the 1948 Olympics, it was one of the first dinghies to be mass-produced at Fairey Marine, by a method of hot-moulding developed during the war for building Mosquito aircraft. The first fibreglass sailing boat in the world had been produced in the United States by Carl Beetle and exhibited at the 1947 New York Boat Show, but it was a chance meeting on the Isle of Wight between VT Halmatic's founder Patrick de Laszlo and Uffa Fox that led to the manufacture in 1952 of the first British fibreglass sailing boat – a Flying Twenty. This was soon followed by a new class of production sailing dinghy – the Flying Fifteen. Another of Fox's designs, this was the first British class of planing keelboats. A prolific author, Fox's

The Uffa Fox-designed Flying Twenty, which became Britain's first fibreglass sailing boat in 1952. (VT Halmatic)

many books inspired a new postwar generation of dinghy designers to further widen the appeal of the planing concept. In his later years Uffa Fox designed the 18ft Jollyboat class, which he assumed would be more tranquil to sail than the Fourteen. In fact it became his fastest dinghy. In 1954, sailing off Cowes in a Force 5, a Jollyboat covered 5 cables at 13.4 knots, which seemed then to be a phenomenal speed.

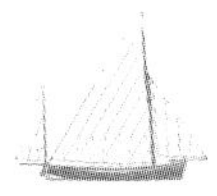

Avenger is preserved at the Cowes Maritime Museum, Isle of Wight. See also the website of the International 14 class, www.i14.org; and Uffa Fox Ltd, www.uffafox.com.

John Illingworth, Ocean Racer

The simple, efficient features that characterise the appearance of the latest sailing cruisers today – vertical stem, flat decks, and high freeboard – were considered sensational in 1947 when they first appeared on a revolutionary offshore racer called *Myth of Malham*. Her creator was John Illingworth RN, an innovative engineering officer who, finding himself stationed in New South Wales when the war ended in 1945, bought an old canoe-sterned 30ft yacht called *Rani*. After giving a lecture to a Sydney boat club, he was invited to join a Christmas-time cruise in company to Tasmania, but Illingworth proposed making a race of it. The distance was about the same as the Fastnet race. *Rani* was the smallest yacht in the competition, and when her radio broke in atrocious weather conditions Australian newspapers reported that she had foundered. But Illingworth had prepared the old cutter meticulously and, driving her just hard enough, famously won the event – the first ever Sydney–Hobart race.

Illingworth had narrowly failed to win the 1937 Fastnet race in his 48ft sloop *Maid of Malham*. At the time it was unthinkable that

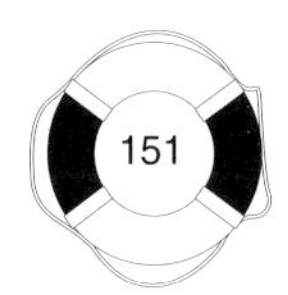

offshore yachts should not have an overhanging spoon bow and counter, and yet under the existing principal Royal Ocean Racing Club (RORC) rule – which was virtually an overall length rule – such overhangs were very heavily penalised. During the war years Illingworth had begun to realise that significant loopholes in the rules could be exploited, and he wondered if the overhangs were structurally necessary to achieve a dazzling windward performance? Returning to England he phoned yacht designer 'Jack' Laurent Giles and revealed his intention to build a radical new 33ft waterline length cutter called *Myth of Malham* to 'go flat out for the Fastnet'.

A progressive and forward-looking designer, Laurent Giles was the ideal character to put into practice Illingworth's detailed written specifications. In addition to having snubbed ends rather than conventional overhangs, Illingworth intended *Myth of Malham* to be innovative in several other respects. To exploit another loophole her freeboard was made high to give large measurements from the deck to the hull, thus confusing the officials into thinking the yacht had a big displacement. In fact Illingworth's instructions for a fairly lightweight triple-skin cold-moulded mahogany hull enabled her displacement to be only just 7.6 tons – which in those years was remarkably light. This allowed Laurent Giles to create a radically innovative shape below the waterline. All other offshore racers still had conventional long keels but on *Myth of Malham* it was found to be possible to achieve further weight-saving by having an underwater profile with a long instep, and an unusually short, deep keel (giving a 7ft draught).

Another loophole Illingworth found in the RORC rules enabled a radical innovation to be made in terms of the sail plan. The foretriangle was cheaply rated so Illingworth made it large and the mainsail small. Already something of a rigging expert (in 1928 he had pioneered the Bermuda masthead rig), he went a stage further, introducing a further innovation by stepping *Myth of Malham*'s mast directly in her centre. Below decks he used his engineering know-how as a prewar designer of submarines to get the most from every cubic foot of space, and to exercise several new ideas. Incorporated into the design were lightweight gadgets for standing and running gear and deck fittings, which Laurent Giles had designed to keep weight to a minimum. With an eye to saving weight, Giles went to the lengths of designing special stay tensioners and even winches; he also introduced mainsheet and headsail sheet lead tracks, with adjustable spring-loaded plungers, and used new alloys for those tracks, which gave unprecedented control over sail trim.

'You ugly little boat', competitors shouted across at *Myth of Malham* at the start of the 1947 Fastnet. They had reason to be concerned. Launched that July, *Myth* had already won her debut blue ribbon event, the prestigious Channel Race, finishing 16 hours ahead of her nearest rival. In the Fastnet the larger offshore racers were well ahead until the turn to windward at the Scilly Isles, when *Myth* really came alive; she danced through the light seas, tacking through about 82 degrees. Rounding the Fastnet, Illingworth hoisted a spinnaker which speeded her to Plymouth and her first Fastnet win. Within a fortnight came another victory in the Plymouth–La Rochelle race. After her hat-trick of historic wins Illingworth shipped her across the Atlantic to America – where she

FASTEST SAILCRAFT

SPEED	TYPE	NAME	WHERE	WHEN
143mph	Ice yacht	Debutante	Wisconsin, USA	1938
116.7mph	Land yacht	Iron Duck	Nevada, USA	1999
46.82knots	Sailboard	F2	Saintes Maries, France	2004
46.52knots	Sailboat	Yellow Pages	Victoria, Australia	1993
41.67knots	Kiteboard	North Rhino 05	Walvis Bay, Namibia	2004

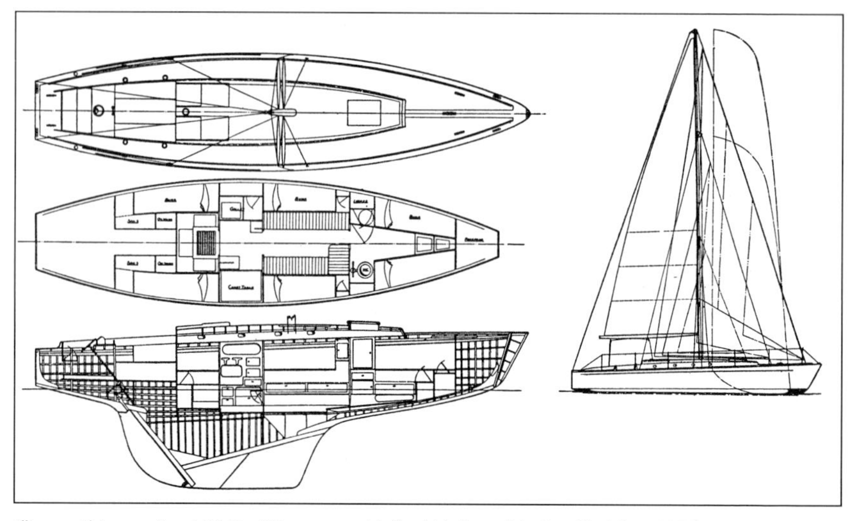

Illingworth's sensational 1947 offshore racer, *Myth of Malham*. (Heaton, *Yachting*, 1955)

promptly won both the famous Long Island Sound races. Next he entered her in the Newport–Bermuda race, and did well, despite the yacht's unfavourable rating under American racing rules. In 1949 *Myth of Malham* won her second Fastnet.

Myth had already become more than a myth. She was almost a legend, being nearly unbeatable on the water. Illingworth's revolutionary, if ugly, creation had successfully broken the cardinal rule of sound experimentation – that only one novelty should be tried at a time – and shattered the age-old maxim that 'A boat which looks right *is* right'.

A larger than life character, the multi-talented Illingworth had huge natural vitality and lived life at a frantic pace. Twice-married, his passion for sailing was funded by his family's estate at Malhamdale in Yorkshire. Abandoning the expensive sport of polo, he owned a string of racing yachts, eight of which, true to his origins, had 'Malham' in their names. A fiercely competitive racer, Illingworth was also a big-hearted character who selflessly used his privileged position to encourage others to advance themselves in the world of sailing. The kudos he gained with *Myth of Malham* was such that in 1948 he became Commodore of the RORC. That year Captain Illingworth began writing a yachting book called *Offshore*. It was one more addition to his phenomenal workload. At the time he was commodore of the Royal Naval Sailing Association (RNSA), he worked at the Admiralty as deputy director of coastal forces, and was in the process of establishing the Sail Training Association. *Offshore* revealed the detailed secrets of *Myth of Malham* – her race strategies, her sail and rigging, her construction and sophisticated gadgetry – and also emphasised Illingworth's humane concern for crew well-being and safety. Reprinted many times in subsequent decades, *Offshore* became the bible for the development of modern ocean racing.

Undeterred by the RORC's scepticism of his belief that very small yachts could be perfectly safe ocean racers, in 1951 Illingworth helped found a new club, the Junior Offshore Group, which sailed 16ft to 24ft midget racers offshore. Several such racers were built at Aero Marine, the yard he owned at Emsworth. Also built there was a 24ft overall length version of *Myth of Malham* called *Wista*, in which he personally won every Junior Offshore Group race entered in 1954, reaffirming his idea that even at that scale short overhangs and fin keels with separate skeg rudders worked perfectly well offshore. However, the radical configuration initially met with sales resistance, being literally a decade ahead of its time.

After leaving the Royal Navy and Aero Marine in 1955, Illingworth formed a business partnership with Angus Primrose. Among their many designs were the rig conversion of *Bloodhound* for Prince Philip; *Gipsy Moth IV*, in which Francis Chichester sailed alone around the world; and, *Oryx*, which Illingworth himself skippered for the 1969 French Admiral's Cup team. The Admiral's Cup itself was Illingworth's idea and was first awarded in 1957.

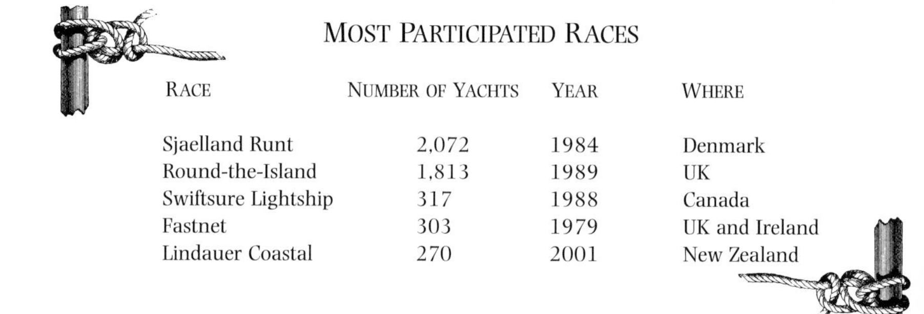

MOST PARTICIPATED RACES

RACE	NUMBER OF YACHTS	YEAR	WHERE
Sjaelland Runt	2,072	1984	Denmark
Round-the-Island	1,813	1989	UK
Swiftsure Lightship	317	1988	Canada
Fastnet	303	1979	UK and Ireland
Lindauer Coastal	270	2001	New Zealand

American yachts would come over specially for the Fastnet but also sailed for good measure in the Channel Race and the main Cowes Week events, so he suggested those races should be combined for a series prize. Entirely coincidentally, in 1957 Illingworth raced *Myth of Malham* in the Class III section of the Fastnet, and ten years after her initial victory she won yet again.

By then *Myth of Malham*'s revolutionary design had been acknowledged as a quantum leap forward in yacht design, just as *Jullanar* and *Gloriana* had been in the nineteenth century. As for Illingworth, he died in 1980, widely respected as the founding father of modern ocean racing.

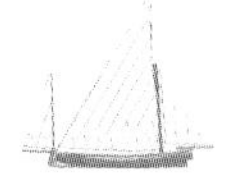

In 1972 *Myth of Malham* was lost in a storm, but each year the Royal Ocean Racing Club organises the 'Myth of Malham Race', www.rorc.org. See also www.laurentgiles.co.uk.

6

Engines, Steering and Anchors

John Hawkins's Steering Wheel

The steering wheel began appearing on European ships in the earliest years of the eighteenth century: in England in 1704, Denmark in 1708 and France in 1709. The Admiralty seem to have been remarkably quick to fit Royal Navy ships with steering wheels as it was only in the previous few years that the Navy Board had begun examining working models of ships that were intended to be equipped with the new-fangled device. One such is a model of a two-decker English warship of 50 or 60 guns. Curiously it had a hutch on the poop deck to protect the head of the helmsman, who would steer from that position by turning a wheel whose barrel would accordingly shorten or slacken the tiller lines attached to the tiller.

According to an academic called D.W. Waters, in 1985 evidence came to light indicating that the Admiralty steering wheel first proved itself to be devastatingly effective in August 1704 at the Battle of Malaga. To seal the conquest of Gibraltar, it was thought necessary for Britain's Admiral Rooke to defeat a French fleet counter-attacking off Malaga. This Rooke did by maintaining a formal line and by careful spacing of the ships throughout the battle – a feat seemingly made possible by the new steering wheels which enabled the English helmsmen to see the foot of their lower

Admiral John Hawkins became the unwitting originator of the steering wheel by innovating the continuous chain pump, which was then adapted into an early steering mechanism. (Holland, *Heroologia*, 1620)

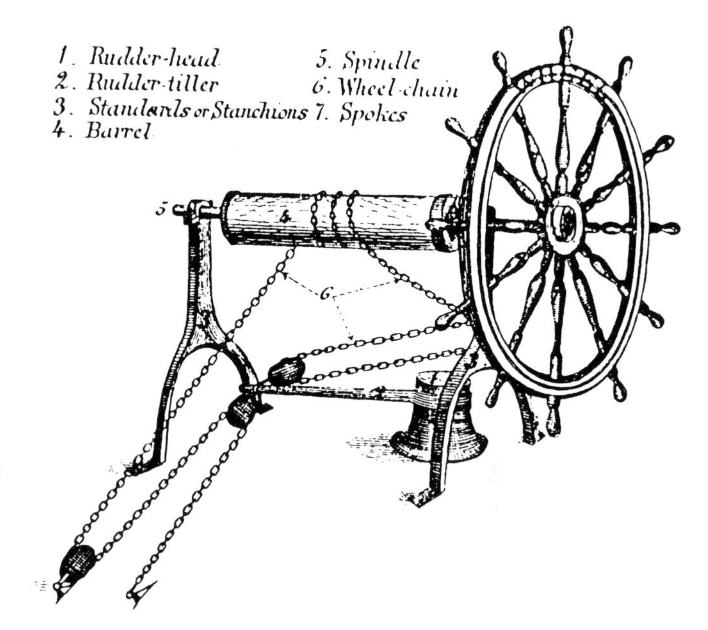

An early chain and wheel system. (Paasch, *Illustrated Marine Encyclopedia*, 1890)

sails and thus see how they drew. The wheel itself, in its most primitive form, was believed to have been invented by the Assyrians in ancient times. Its earliest use, which eventually became relevant to steering, was as a part of the endless chain pump invented by Admiral Sir John Hawkins in the 1580s, and the wheel has been in use ever since. It is not known who first thought of using this contraption as a steering mechanism, with its wheel of ten radical spokes, but it does seem that the actual machine for steering was derived from the chain pump invented by Hawkins.

The National Maritime Museum, Greenwich, has models of ships fitted with the earliest steering wheels, www.nmm.ac.uk.

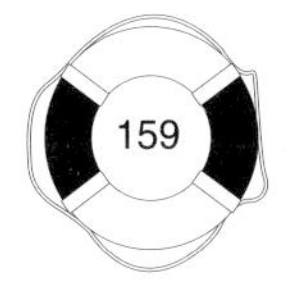

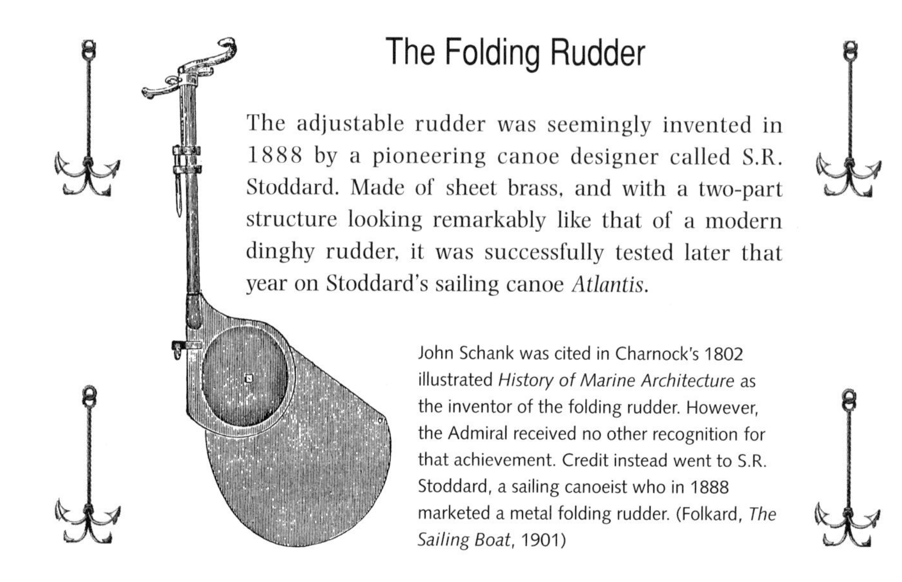

The Folding Rudder

The adjustable rudder was seemingly invented in 1888 by a pioneering canoe designer called S.R. Stoddard. Made of sheet brass, and with a two-part structure looking remarkably like that of a modern dinghy rudder, it was successfully tested later that year on Stoddard's sailing canoe *Atlantis*.

John Schank was cited in Charnock's 1802 illustrated *History of Marine Architecture* as the inventor of the folding rudder. However, the Admiral received no other recognition for that achievement. Credit instead went to S.R. Stoddard, a sailing canoeist who in 1888 marketed a metal folding rudder. (Folkard, *The Sailing Boat*, 1901)

Andrew Smith's Wire Rigging

An engineer called Andrew Smith, having in 1829 devised a radical new type of window shutter that could be closed from inside a building, needed to find a sash cable of weather-resistant quality to operate the new contraption. He reckoned what would do best was a wire cable. Soon realising that this could have nautical applications, Smith began experimenting with various means of applying wire ropes to ships' rigging, then in 1835 successfully obtained a patent for 'A new standing rigging for ships and vessels and a new method of using it.' He manufactured several kinds of wire rope for that purpose, using the ropework techniques of the hemp cordage industry, and by the early 1840s British naval and merchant ships were already making much use of his 'metal cordage' wire rope for standing rigging.

Smith had been fortunate to get his invention established fractionally before a German called Wilhelm Albert, who for many years had been attempting to make cables from twisted lengths of wrought-iron wire. Those so-called 'Albert ropes' were originally developed for operation in the mineshafts of the Harz silver mines of north Germany, where the incessantly damp conditions were inclement for traditional chains and hemp cables.

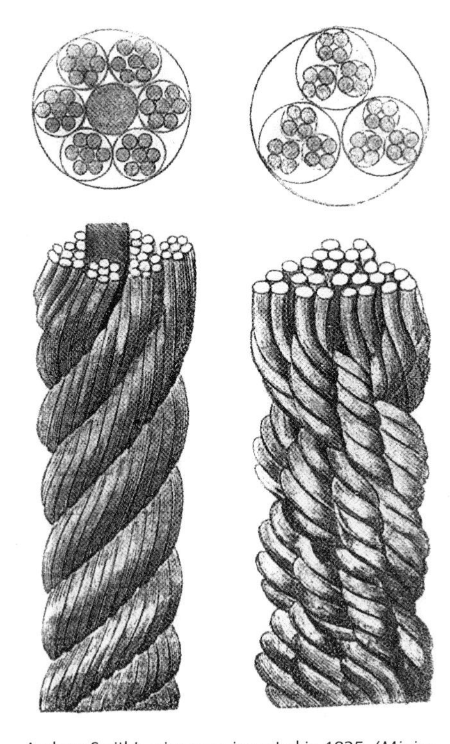

Andrew Smith's wire rope, invented in 1835. (*Mining Journal*, December 1844)

However, a serious dispute arose in 1840 when a new rapid transit system known as the Blackwall railway opened for business in London using Smith's wire ropes instead of hemp haulage. In the meantime another inventor, Robert Newall, having learnt about the Albert ropes, devised a means to make wire ropes in a factory using machinery rather than the hand-twisting method. His ropes were tested with success in the Blackwall railway, but Smith opposed Newall's efforts during a patent fight in the 1840s, in which Newall eventually prevailed. Shaken by the litigation, Smith left for America and the Gold Rush. In America the wire ropes Smith had pioneered became hugely popular. And in England, a nautical writer called Edwin Brett was able to note in 1869 in his books *Notes on Yachts*, 'Wire rope is now in general use for the standing rigging of yachts.'

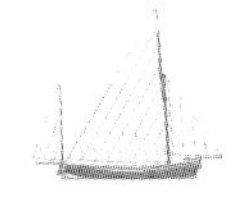

The Ropery at Chatham Historic Dockyard, Kent, www.chdt.org.uk.

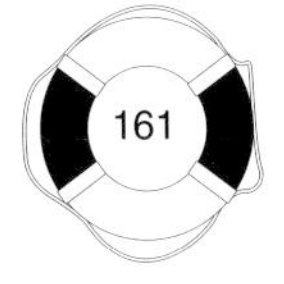

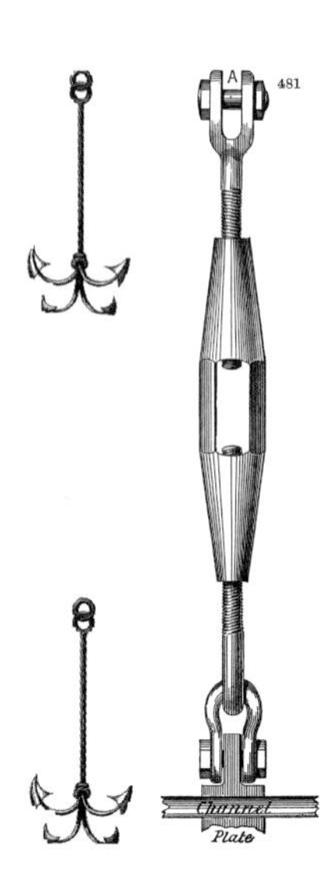

The Turnbuckle

Originally invented in 1470 by an Italian Renaissance artist called Francesco di Giorgio as a dental instrument for the purposes of strengthening teeth, the turnbuckle was reinvented in the 1870s as a means of tensioning telegraph wires, and thence came to be used afloat as a rigging screw.

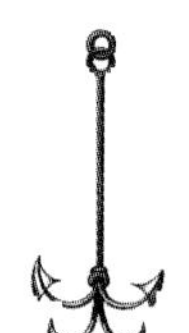

The turnbuckle was originally developed for dentists. (Dixon Kemp, *Yacht Architecture*, 1891)

Francis Smith's Screw Propeller

In February 1837 a steam launch puffing along the Limehouse Canal near the London Docks inadvertently clipped an object in the water which snapped off nearly half of her 15in-long experimental wooden propeller. Since various top-hatted dignitaries were aboard at the time the collision could have been a publicity disaster for the 34ft craft, which was struggling to prove the effectiveness of its two-complete-turn propeller (similar in appearance to fusilli pasta). But amazingly the boat instantly went much faster! Quite by accident a more efficient shape for the screw propeller had been discovered.

Curiously the inventor of this revolutionary propeller was a farmer. For no apparent reason it was on his parents' Romney Marsh sheep farm that young Francis Pettit Smith developed a prodigious skill in the construction of model boats, and devised ingenious methods of propulsion for them. In 1834, while he was a self-employed tenant farmer at Hendon, just north-west of London, he constructed a 2ft model boat propelled by a screw driven by a spring; it moved so well that he realised this could be preferable to paddles as a propulsion device for vessels. Persuaded by friends who saw him experimenting with the model in the farm's horsepond, he exhibited it to the public at the Adelaide Gallery in London, where it was seen by the prominent banker John Wright. Impressed, he offered to finance its development. Reckoning he would soon be financially secure, Smith virtually abandoned agriculture and on 31 May 1836 took out a patent 'for propelling vessels by means of a screw revolving beneath the water'. Subsequently he oversaw the

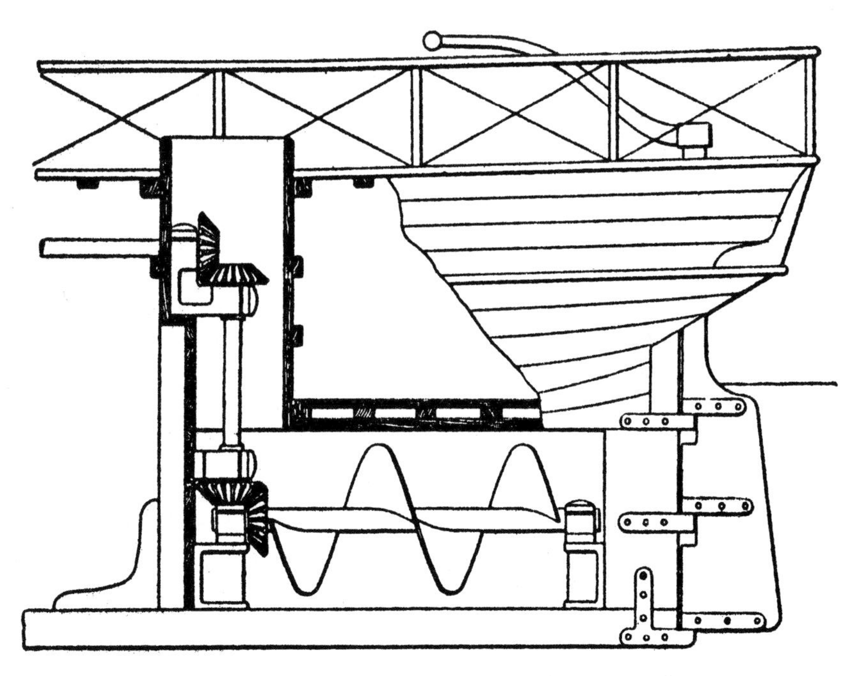

In 1837 Smith discovered that when part of this experimental screw propeller arrangement was accidentally damaged the trial boat went faster! (Seaton, *The Screw Propeller*, 1909)

Francis 'Pettit' Smith, the sheep farmer who created Britain's first successful propeller. (National Portrait Gallery)

construction of an experimental launch called *F.P. Smith*. In September 1837, with her 6-in diameter metal propeller reduced to just a single complete turn, Smith took the launch on a voyage round the Kent coast in choppy water, thus proving the simple screw was appropriate for sea voyages.

At this time paddle-steamers were the only mechanised vessels but they had serious disadvantages. When they rolled in a heavy sea, one paddle-wheel would be lifted out of the water while the other was deeply submerged, thus risking a broken paddle-shaft and putting a dangerous strain on the engine. However, since paddle-wheels were already so widely in use, Smith came up against the commercial reality that shipowners were unlikely to be readily impressed by his experimental launch. Especially because since 1681, when the feasibility of using a screw as motive power for ships was initially considered by Robert Hooke, many absurd devices had been proposed. In 1802 a 'perpetual sculling machine', patented by Edward Shorter, was tried out on a merchant ship, but it was unsuccessful, being driven by an arrangement of ropes and pulleys operated from a crew-manned capstan. It was only in 1832 that a French experimentalist, Frédéric Sauvage, became the first European to hit upon the idea of applying steam power to the screw. Meanwhile in America a screw-driven boat had been created in 1804 by Colonel John Stevens (father of the New York Yacht Club's eventual founder), but it was not much developed. A steam-driven system devised in 1836 by the Swedish engineer John Ericsson had two propellers revolving in opposite directions directly aft of a rudder, but this was considered too hazardous for steering. In the absence of any tank-testing, all such experiments were somewhat hit-and-miss.

Crucially, unlike the other propeller pioneers, Smith's research had so far been well funded. But as yet he had received no income from the sales of his invention. He reckoned he would need the Admiralty's acceptance of the propeller before cautious commercial shipowners would order vessels to be so equipped. However, the Admiralty, being mostly interested in finished products ready for service, expected him to provide a full-sized demonstration vessel for extended trials at his own expense. So Smith established the Ship Propeller Company, funded by an impressive syndicate of notables, including the keen yachtsman Lord Sligo, Governor of Jamaica. (Interestingly, Smith's father had once been Lord Sligo's classics tutor.)

The 237-ton *Archimedes* built at Wapping in 1839 became the first sea-going propeller-driven yacht, and indeed was the prototype of the screw schooner that was to dominate steam yachting for almost a century. At 10 knots she could do more than double the Admiralty's required speed, and she was sailed as far afield as Portugal, but even though the Admiralty chartered her for large-scale tests they still would not place orders. However, in Bristol, during a much-publicised circumnavigation of Britain in 1840, the revolutionary yacht made an important ally. After inspecting her, the legendary engineer Isambard Kingdom Brunel decided to convert the 1,320-ton transatlantic paddle-steamer SS *Great Britain* he was building there into a propeller ship. That spurred the Admiralty into action, and they allowed Brunel to participate in a similar propeller-conversion of one of their ships, the 1,050-ton *Rattler*. Brunel effectively took over the development of the screw between 1840 and 1844, and with the two projects running concurrently, he scientifically tested thirty forms of screw – from which he chose Smith's as the best.

Finally, in April 1845 the Admiralty sought to test the virtues of Smith's device by means of a spectacular propeller versus paddle tug-of-war off the Nore in the Thames estuary. HMS *Rattler* and HMS *Alecto*, an identically sized paddle-sloop, were tied stern to stern and ordered to steam at full speed ahead. Gripping the water better *Rattler* first held her position and then began to tow *Alecto* backwards at a speed of 2.8 knots. So impressed were the Admiralty officials by this conclusive test that they accepted the propeller, and by 1854 the official Navy List included the names of ninety-seven screw vessels. All the while efficient refinements to Smith's original discovery were being spawned by other innovators, producing devices such as a feathering propeller, a variable pitch propeller, a propeller shaft clutch and even a folding propeller. An additional consequence of the acceptance of the screw was the widespread appearance of steam launches, many of them built by Alfred Yarrow or by John Isaacs Thornycroft (who in 1862 produced the first screw launch quick enough to keep up with the Boat Race eights).

But Francis Smith did not fare as well as he had hoped from his invention. Although he had been employed until 1850 advising the Admiralty on the installation of every propeller in their ships – and thus helped to maintain Britain's naval supremacy – he had been paid a virtual pittance for it. To add insult to injury, in 1855, just before his patent expired, a full and final £20,000 prize was awarded for the invention of the propeller but Smith received just a fifth of it, the remainder going to rivals he believed his achievements had seen off long ago. Unable now even to cover the running costs of the demonstration yacht *Archimedes*, his Ship Propeller Company collapsed. In despair Smith fled to Guernsey where he unsuccessfully reverted to farming. But he was compelled by pecuniary needs to accept the post of curator of the Patent Office Museum in South Kensington. The only acknowledgement of his services to marine invention was the conferment on him of a knighthood three years before his death in 1874.

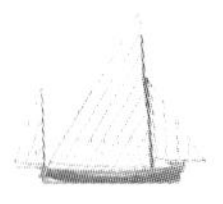

Models of the SS *F.P. Smith* and others showing the evolution of propellers are at the Science Museum in London, as is HMS *Rattler*'s original propeller. The SS *Great Britain* is preserved at Bristol, www.ss-great-britain.com.

The Central Control Cockpit

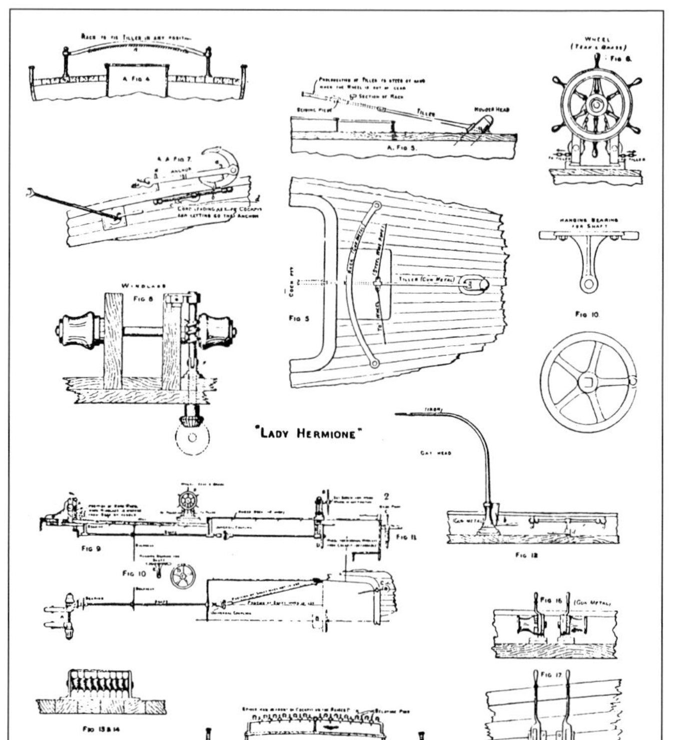

When built in 1881, Dufferin's yawl *Lady Hermione* had various ingenious devices to enable her to be sailed single-handed. (Pritchett *et al.*, *Yachting*, 1894)

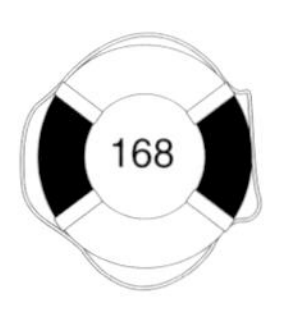

The Central Control Cockpit

The first yacht with all its lines leading to the cockpit (see facing page) belonged to Lord Dufferin, who had served as Viceroy of India from 1884 to 1888. During the early 1890s he was the British Ambassador in Paris. During this time, for the purposes of single-handed sailing, he built *Lady Hermione*, a 5-ton yawl on which all the lines led to the cockpit though special deck-mounted oblong blocks. He could even raise and lower the anchor without leaving the cockpit. A keen yachtsman, Dufferin was best known as the author of the popular cruising account *Letters from High Latitudes* (1857).

Nathaniel Herreshoff and Self-steering

Until the 1930s it had generally been assumed that the idea of self-steering had originated with the enthusiastic model-boat sailors in the Round Pond in Kensington Gardens, who had seen the need for something better than a weighted rudder and a well-balanced sail plan. In 1904 George Braine developed a new device to steer model boats, which consisted of a quadrant attached to the tiller and a rubber band! It was, however, only an improved method of using the sails to control the steering. The first really effective wind vane was used by a Norwegian called Sam O. Berge, who won a model boat international championship in 1935 with *Prince Charming*, a model yacht fitted with a device controlled by a true wind vane. In the *Model Yachtsman* magazine that year the ebullient Berge published his design and talked of patenting it, but a rebuke was swiftly issued by a designer called Captain Nathaniel Herreshoff, who claimed he had invented the self-steering vane back in 1875!

Bond Movie Villain Boats

Boat	Chief villain	Movie
Disco Valante hydrofoil	Emilio Largo	*Thunderball* (1965)
Bathosub	Ernst Stavro Blofeld	*Diamonds Are Forever* (1971)
Chinese junk	Scaramanga	*The Man with the Golden Gun* (1974)
Sleath catamaran	Elliot Carver	*Tomorrow Never Dies* (1997)
Sunseeker Superhawk 34	Renard	*The World is Not Enough* (1999)

The basic principle of the Herreshoff gear was even simpler than that of Braine's gear: the vane 'read' the wind direction over the stern of the boat and moved the rudder if the wind direction changed. Known as the 'Wizard of Bristol', Herreshoff dominated American yacht racing for seventy years, and indeed was the greatest American yacht designer ever. *Gloriana*, the long keel 46-footer with a deep cutaway profile he designed in 1891, became the prototype of the modern yacht. Originally he had specialised in building fast steam launches for wealthy customers at his yard at Bristol, where he pioneered various labour-saving techniques – such as building boats with the keel uppermost before turning them over for fitting out. This later became standard practice. In the 1890s he switched to building sailing craft and went on to construct several innovative America's Cup defenders.

In 1936, a year after Herreshoff's original invention became public knowledge, a Frenchman called Marin-Marie fitted a similar device to a motor yacht called *Arielle*, which he then successfully took across the Atlantic, thus establishing the seaworthiness of wind vane self-steering.

Herreshoff Marine Museum, Bristol, Rhode Island, www.herreshoff.org.

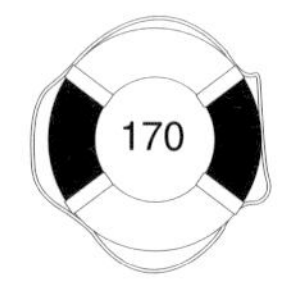

Rudolf Diesel's Boat Engine

The first petrol-engined boat was the speedy 23ft *Die Sieben Schwaben* ('The Seven Swabians'), powered by a V-twin-cylinder, which appeared on the River Elbe in 1888. The creation of Adolf Daimler, she made a profound impression, despite only having a 2hp engine. Convenient and light though petrol engines were (compared to steam engines), to yacht designers it was evident early on that they suffered from the serious disadvantage that they needed electrical equipment for ignition – which could be potentially dangerous in a marine environment. There was also the risk of highly inflammable gases leaking into the bilges and causing disastrous explosions.

Dr Rudolf Diesel's alternative idea was to have an engine which ran on pulverised coal! That was what he stated on his 1894 patent, although there is no evidence to suggest the engine actually ever ran on coal. Nevertheless, his radically new engine dispensed with the need for a carburettor, the fuel oil being injected and simultaneously vaporised by a blast of high-pressure air through an atomiser in the nozzle and ignited by the heat of compression alone. This was far safer than the petrol engine, since the diesel engine used a fuel with a high flashpoint and it required no naked flame or electric spark to keep it running.

The first marine diesel engine on record was a compact two-cylinder 20hp unit fitted to a barge called *Petit-Pierre*, which was constructed in 1903 for use on the French canals. Despite the potential advantages of diesel engines for yachts, for many years few such engines were made, and in Britain it was not until 1932 that the first RNLI lifeboat was fitted with a diesel engine.

Early examples of Rudolf Diesel's engines are at the Maritime Museum, Hamburg, www.hamburgmuseum.de.

The Anchor Weight

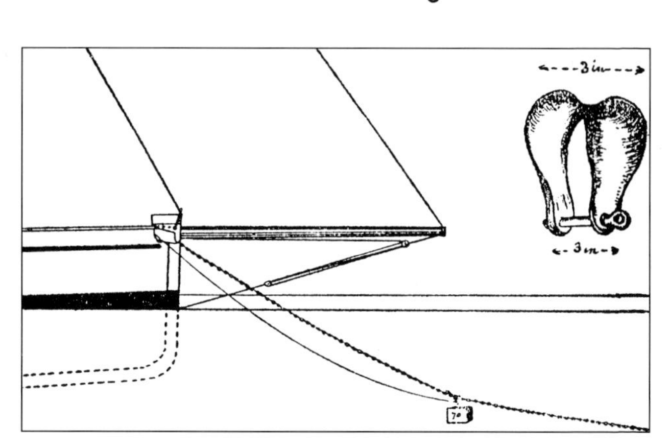

Claud Worth invented the anchor weight . . . (Worth, *Yacht Cruising*, 1910)

Sometimes called an 'anchor angel', the anchor weight was invented by Dr Claud Worth. When his boat was anchored in the Thames estuary during a strong breeze in 1890, he improvised a means of reducing the strain on his fully paid out 40 fathoms of anchor warp. His book *Yacht Cruising* (1910) recounted how: 'We got up a 70-pound pig of lead ballast and tied a rope round it and made a becket round the warp. Then we made fast a small line to the pig. We let the pig slide down the warp, so the weight of the lead pig acted as a kind of spring on the warp.' Worth also introduced the brilliantly simple idea of fitting a ratchet to a stemhead roller to facilitate hauling in an anchor.

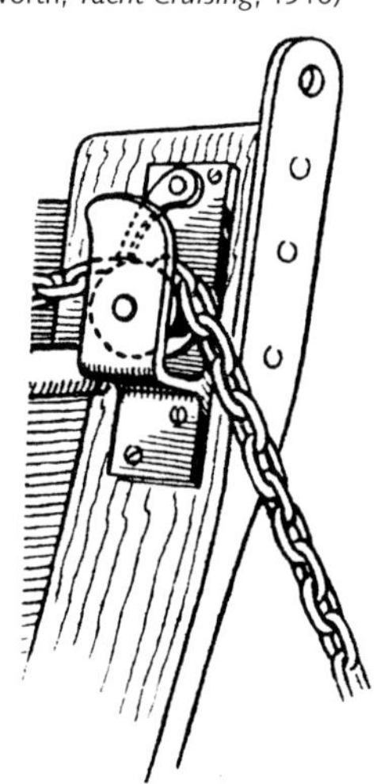
. . . and then a stemhead ratchet mechanism. (*Yachting World*, May 1950)

Cameron Waterman's Outboard Motor

The word 'outboard' was coined by the American Cameron Waterman, although outboards had been used before his time. One of the earliest such devices was an *electric* outboard, designed by the Englishman Ernest du Boulay on the Isle of Wight. Cameron Waterman's version was rather more practical. A law student at Yale, in September 1903 he acquired a Curtis motorcycle. He removed the engine and hung it over the back of a chair in his

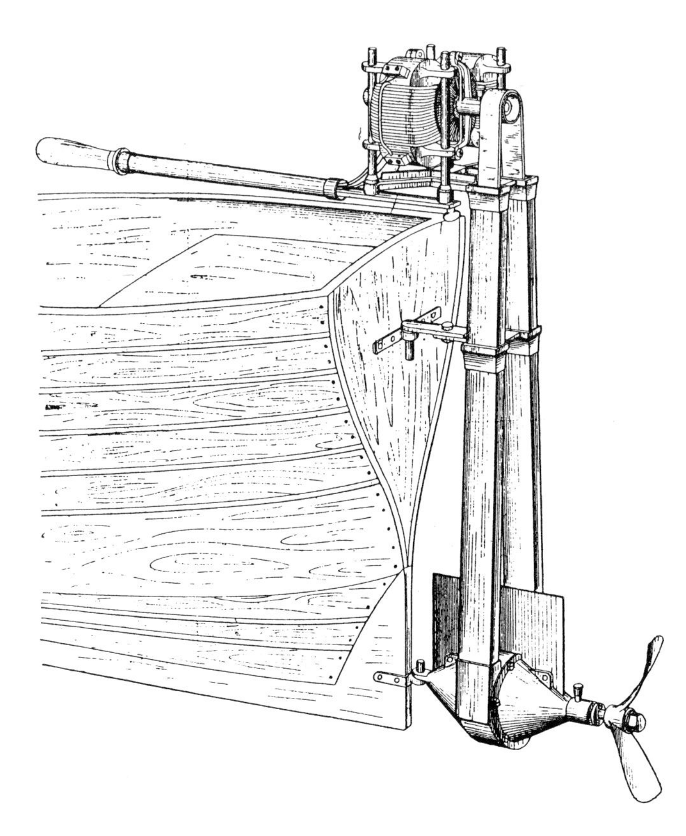

Cameron Waterman's 1903 invention was preceded by a British outboard – powered by an electric motor! (Ernest du Boulay, *A Textbook on Marine Engines*, 1902)

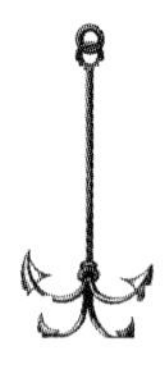

The Drogue Sea Anchor

The yacht sea anchor was famously first used by John Voss to ride out storms during his voyage across the Pacific in the 38ft, schooner-rigged *Tilikum*, a converted Native American war canoe. The nautical writer Edward Knight had initially advised yachtsmen to develop such equipment in 1895. Voss became a celebrity upon reaching England, and eventually publicised the windsock-shaped drogue anchor in his autobiography, *Venturesome Voyages of Captain Voss* (1913).

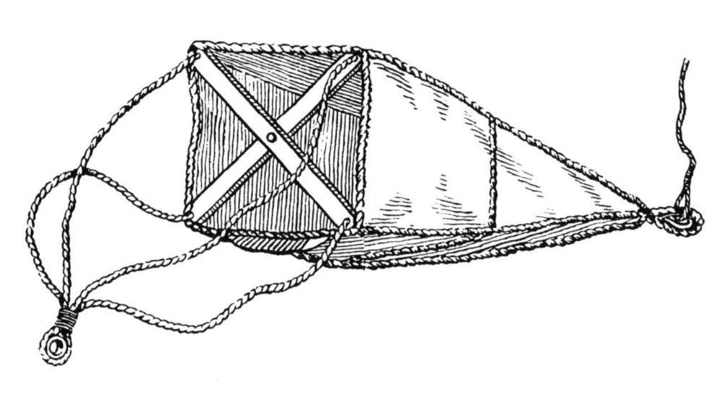

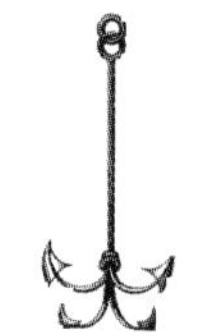

Sea anchor, innovated by John Voss. (Cooke, *Hints, Tips and Gadgets*, 1939)

room in order to overhaul it. A four-cycle, air-cooled motor, it weighed about 20lb. It occurred to Waterman that if he could hang it on the transom of a rowing boat, and attach a propeller to it, it could drive a boat. And if he hinged the engine to the back of the boat it could be used to steer as well. A petrol tank could be mounted near the tiller to make the whole unit self-sufficient. One final idea was to allow the engine to tilt up to a horizontal position to protect it in the absence of a keel or skeg.

In February 1905 Waterman and his father took their working model to the Detroit River and attached it to a 15ft steel rowing boat. Although the river was full of drift-ice, the trial was a complete success

– and it was on that day that the term 'outboard motor' was coined. Commercial production started in 1906 as the 'Waterman Porto' motor, and the engine was exhibited later that year at the National Boat Show in the old Madison Square Garden. The first mass-produced outboard, it marked the beginning of a whole new industry.

Motorboat Museum, Basildon, Essex, www.motorboatmuseum.org.uk.

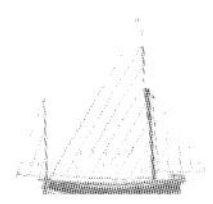

Geoffrey Taylor's CQR Anchor

The CQR anchor – sometimes claimed to mean 'Can be Quickly Released' – was invented by a brilliant mathematical physicist of world renown called Professor Sir Geoffrey Taylor. He was also an accomplished yachtsman, who in 1927 won the Royal Cruising Club's Challenge Cup for a trip he made to the Arctic Circle in his 19-ton cutter *Frolic*. A scientist with an extraordinary ability to concentrate, he would sometimes anchor for the night when cruising and hastily write a paper for the Royal Society!

Invented in 1933, the CQR was originally intended to be used for anchoring seaplanes! (Cooke, *Hints, Tips and Gadgets*, 1939)

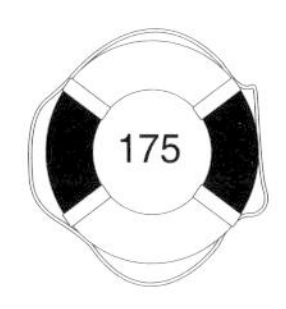

It was partly because of his knowledge of small boats that he was encouraged by a government aviation committee to design a simple but effective new anchor for flying boats. The result in 1934 was the prototype of the plough-shaped CQR anchor, which Taylor patented and subsequently tested with enormous success on a chartered yacht in the Shetland Islands. He found the new 50lb anchor had four times the holding power of *Frolic*'s traditional Fisherman-type anchor. When tested at the Seaplane Experimental Station at Felixstowe the Taylor device would not drag until pulled with sixty times its own weight. During the Second World War some 45,000 CQRs were ordered by the Admiralty for service in small craft. Many of those anchors flooded on to the surplus market in the postwar years, enabling yachtsman to purchase them inexpensively. The Taylor CQR set a new trend for plough anchors and there eventually came to be made several variants of his original design.

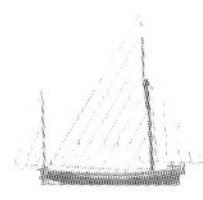

Various anchors are exhibited at the National Maritime Museum, Greenwich, www.nmm.ac.uk.

7

Emergencies

Lionel Lukin's Lifeboat

Improbable though it seems, the first Briton systematically to devote his energies to readying boats to save lives was a Master of the Worshipful Company of Coachbuilders, who, as such, was more at home selling shiny vehicles to the gentry at his fashionable premises in Long Acre, London. Although a landsman, coachbuilder Lionel Lukin had the sea in his blood, being descended on his mother's side from Lionel Lane, one of the famous Admiral Blake's captains. He later claimed that it was in 1784, having somehow learnt 'that by the oversetting and sinking of both sailing and rowing boats many valuable lives were being lost', that he set about formulating a means of making boats safer. Lukin was well known in court circles, and the Prince of Wales (later King George IV), who knew Lukin personally, not only encouraged him to test his invention, but offered to finance the whole cost of those experiments.

Rather than specifically designing a boat that would go to the rescue of others, Lukin initially concentrated on figuring out how to make existing boats safer for their occupants. In 1784 he purchased a Norwegian yawl, which he equipped according to his plan and tried out on the Thames. To the outside of the yawl's gunwhale he added a hefty belt of cork, some 9in deep amidships and tapering towards the head and stern. Within the boat, from

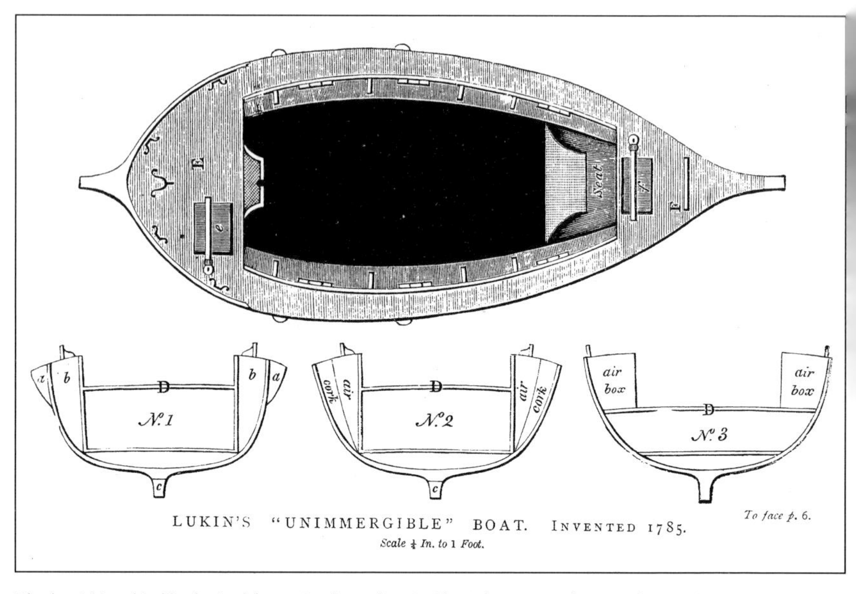

The boat Lionel Lukin devised for saving lives. (Lewis, *The Life-Boat and its Work*, 1874)

the gunwale to the floor, he created hollow watertight chambers, and he also increased the buoyancy of the boat by having watertight tanks at the bow and stern. By all these means he reckoned the total buoyancy would be more than enough to keep the vessel afloat even when completely waterlogged. In order to give the boat sufficient ballast to keep it upright, he added a false iron keel. Lukin constructed several trial boats, all of which he found to be unsinkable, and he took out a patent for the design in November 1785.

It was curious that this coachbuilding entrepreneur, who had even gone so far as to produce a plush sales promotion pamphlet for his radical boat, should have insisted on giving the craft the disquieting name *Unimmergible*! Perhaps not surprisingly, irrespective of the Prince of Wales's potentially influential endorsement, no one was prepared to buy. In vain Lukin appealed to the First Lord of the Admiralty, and to various admirals and captains in the Royal Navy. Even the deputy master at Trinity House would only advise him to lend *Unimmergible* to the Ramsgate pilot for extensive sea trials. That was the last Lukin saw of his beloved prototype. She was soon being used for cross-Channel smuggling trips until she was captured and destroyed by fire. Undeterred Lukin consoled himself by having another boat built, which he called *Witch* on account of her superb sailing qualities in rough weather.

But then in 1786 someone expressed an interest in Lukin's boats. A philanthropic clergyman, Archdeacon Sharp, devised schemes for the benefit of mariners and shipwrecked persons on behalf of a charitable trust at Bamburgh Castle on the Northumberland coast. Sharp sent a fishing coble to the Long Acre workshops for conversion into a safety boat. The necessary enhancements were carried out under Lukin's supervision, and the boat (which Lukin named the *Insubmersible*) was then sent back to Bamburgh and employed for some years in saving lives from shipwrecks. Thus Lukin effectively created the first lifeboat, and it was used to equip the first life-saving station on the coast. Nevertheless with the exception of Bamburgh, there continued to be no enthusiasm for lifeboats. (There was said to have been a lifeboat station in the 1770s near the Mersey approaches, but it is not known if that was equipped with a custom-made lifeboat.)

All that would change in dramatic fashion. On 15 March 1789 a Whitby collier called *Adventure* was wrecked at the mouth of the Tyne, with her crew falling from the rigging to their deaths in the fierce seas in full view of thousands of horrified spectators helpless to render assistance. It was a momentous event. Profoundly shocked, the elders of South Shields organised a competition to select a specially designed lifeboat able to perform rescues in such

THE INVENTION,

PRINCIPLES OF CONSTRUCTION,

AND USES OF

UNIMMERGIBLE BOATS,

STATED IN A LETTER TO HIS ROYAL HIGHNESS

THE PRINCE OF WALES,

BY LIONEL LUKIN.

LONDON:

PRINTED FOR THE AUTHOR,

By J. Nichols and Son, Red Lion Passage, Fleet Street;

AND SOLD BY T. BECKETT, PALL MALL; T. EGERTON, WHITEHALL; AND J. ASPERNE, CORNHILL.

1806.

The Royal coachbuilder's prospectus for the *Unimmergible*. (Lionel Lukin, *Pamphlet*)

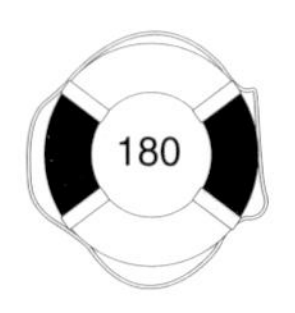

difficult waters. Apparently unaware of this contest, Lukin never got a chance to enter. William Wouldhave, a local house painter, presented what eventually became the self-righting lifeboat (inspired by observing a segment of a sphere floating in a bowl). It was an idea that Wouldhave had figured out back in 1789 but had never patented. The chosen design was put forward by a local boatbuilder called Henry Greathead and was effectively a curved keel surf boat, whose buoyancy derived from flotation tanks and cork blocks identical to those developed by Lukin. A prototype (retrospectively called the *Original*), which he perfected with Wouldhave's help, was approved by the committee and the business began to boom. By 1804 Greathead had already received thirty-one orders for lifeboats for the British coast and abroad. Honours and treasures were heaped upon Greathead in the form of medals, a huge parliamentary grant of £1,200, and prizes from Trinity House and Lloyd's. But Greathead, a Jack-the-Lad character with a talent for self-publicity, went a step too far when he claimed in his promotional literature that he was the 'inventor of the lifeboat'.

Unlike Wouldhave or Greathead, Lukin's interest in lifeboat design was altruistic, and not at all driven by a quest for prestige or financial reward. So when in 1806 a bitter controversy began to rage in the letters pages of the *Gentleman's Magazine* about who was the real originator, Lukin loftily professed to have 'no curiosity to enquire into the local squabbles of a remote country town'. Indeed, he never went further than to note that Greathead had been careful not to collect his prizes until Lukin's patent had expired. Changing tack, Lukin devoted his energies to responding to a plea by the Suffolk Humane Society to develop a *sailing* lifeboat. In 1807 he designed the first of five such vessels. Some 40ft long, and with superb sailing abilities, for many years they saved hundreds of lives off the Norfolk and Suffolk coasts, where they could search the outlying sands as no other boats could.

BUSIEST RNLI LIFEBOAT STATIONS

STATION	NUMBER OF 'SHOUTS' IN A YEAR (2003 FIGURES)
Poole	179
Torbay	133
Southend-on-Sea	125
Weymouth	103
Tenby	78

Having outlived Greathead and Wouldhave, who died in 1816 and 1821 respectively), Lukin retired to Hythe in 1824 – the year in which he had the satisfaction of seeing the establishment of the nationwide lifeboat service that was to become the RNLI. He died at Hythe in 1834. Even though Lukin might justifiably have claimed to have built the first oared lifeboat, equipped the first lifeboat station and created the first sailing lifeboat, Greathead's talent for publicity ensured that Lukin's fine achievements were overshadowed. The only existing acknowledgement to Lukin was in Hythe parish church, where there was a stained-glass window in his memory.

The oldest surviving lifeboat, *Zetland* (an 1802 Greathead), is preserved at the RNLI Zetland Lifeboat Museum, Redcar, Yorkshire.

George Manby's Fire Extinguisher

The inventor of the fire extinguisher would doubtless have been surprised to learn that his fame rests on that device and not on his other life-saving mechanical inventions. Born in 1765, George Manby's original intention was to follow in his father's footsteps and become a captain in the Fusiliers. At the age of seventeen he volunteered to fight in the American War of Independence but was

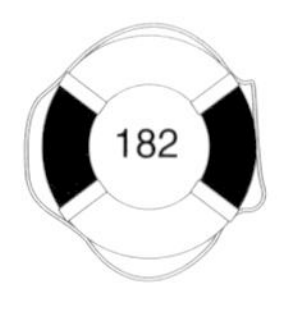

rejected for being too young, so he settled for a commission in the Cambridge militia, until an attack of rheumatism in the knee prevented him from marching and he had to retire to his family estate at Woodhall. The ruination of that property by flooding in 1794 made Manby wonder if his life was jinxed. He moved to Bristol with his wife, and there eked out a living writing travel guides, until a pamphlet he wrote in 1803 condemning Napoleon as a tyrant and arguing for an urgent strengthening of England's coastal batteries, was noted by the Secretary of War, Charles Yorke. That August, Yorke appointed Manby to take charge of the shore defences at Great Yarmouth.

Quite why the Great Yarmouth Barrack-Master should have attended a meeting in the City of London some ten years later on the subject of fire-watching patrols is unclear, except that by then he had apparently invented a 'trampoline' for catching people who

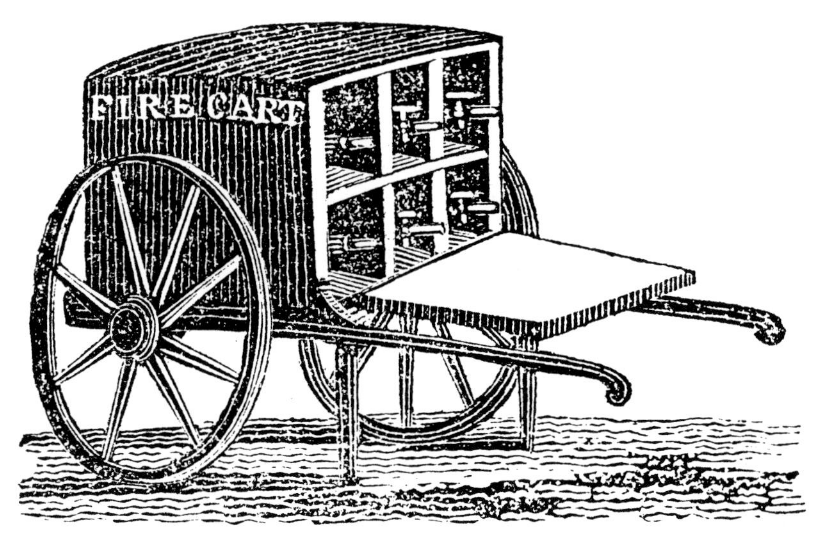

The cart Manby designed for dockyard extinguishers. (Manby, *The Extinction and Prevention of Destructive Fires*, 1830)

jumped from the windows of burning buildings. Reckoning that the City's planned look-out patrols would be more effective if they were equipped with a means of coping with small fires instantly, in 1813 Manby devised a portable high-pressure fire extinguisher! It consisted of a strong copper cylinder, 2ft long and 8in in diameter, with a capacity of 4 gallons. Each extinguisher contained only 3 gallons of water, in which carbonate of potash had been dissolved for the purposes of acting as a flame retardant, while the remaining space was filled with compressed air. When the stop-cock was opened, the air expanded, forcing the water along a tube and out through a nozzle. To facilitate their use, Manby proposed the construction of lightweight 'Firecarts', each of which could accommodate six extinguishers. He even proposed a plan for 'Fire Stations' in different parts of London, where specially trained crews would be instantly ready to tackle fires with the firecarts and extinguishers he had designed.

Regardless of the various successful demonstrations Manby carried out in 1816, no insurance company would finance the prototype extinguisher, or even reimburse him for the development expense. However, later that June a letter arrived from Captain P.B. Pellow RN, who had seen the extinguisher in action and thought that such a device would be ideal for fighting fires on ships. Manby decided to have a go at selling his invention to the Admiralty. Eventually, according to his biographer Kenneth Walthew, a trial demonstration was arranged in Portsmouth before a large crowd of spectators, including the dockyard's superintendent, Admiral Sir Frederick Maitland (on whose ship HMS *Bellerophon* Napoleon had famously surrendered in 1815).

An old sentry box was set ablaze and a Portsmouth dockyard fireman attempted to quell the flames with the extinguisher. At first he squirted the fluid everywhere but in the right direction, convincing Manby that he was deliberately sabotaging the

The extinguisher being demonstrated. (Manby, *The Extinction and Prevention of Destructive Fires*, 1830)

experiment because orthodox firemen were prejudiced against his invention. However, once the jet was aimed correctly, the fire was put out almost immediately. Maitland was impressed. He declared, 'Your firecart, Manby, might be used with advantage in the dockyard as the cylinder it contains can be brought into action for the extinction of fires much sooner than a fire engine.' Manby then awaited a favourable response from the Admiralty.

During a turbulent storm on 18 February 1807 a gun brig called *Snipe* had run aground and broke up just 60 yards off the beach at Great Yarmouth. A helpless witness as sixty-seven of her crew as well as women and children passengers drowned, Manby set about inventing a device to get persons ashore from such dangerously stranded vessels. As a young gunner he had regularly fired a line over Downham Market church, and this gave him an idea. He soon adapted a 6lb mortar to fire a line into the rigging of a stricken vessel. By experimenting Manby improved his idea, discovering that attaching a leather strap to the projectile ball prevented the line snapping under the strain of the blast. Initially he envisaged that the line would be attached to a tailing block and an endless whip line, which would then enable a rescue boat to be hauled between ship and shore.

The first recorded rescue using the new contraption took place in February 1808 when a mortar party commanded by Manby himself saved the crew of the brig *Elizabeth*, wrecked 150 yards off Great Yarmouth. In due course Manby perfected his invention. The recovery boat was replaced by a canvas cot, and then a lightweight mortar was introduced which enabled the entire apparatus to be carried on horseback. At least a hundred sets were manufactured (until superseded by rockets in 1851) and Manby toured the country advising on the mortar's best use, and for his invention he received remuneration of £2,000 from the government and in 1831 was made a Fellow of the Royal Society.

Perhaps Captain Manby should have stopped there. But he had been so traumatised by the sight of that 1807 shipwreck that he was spurred on to produce other inventions; and such was his self-confidence that he was sure they would be financially worthwhile. His subsequent schemes included a device for rescuing persons who had fallen through ice-covered water; a system of kites for transferring written orders between warships in bad weather; an oblong-shaped artillery shell; a harpoon gun for catching whales; and even a plan for using Greenland as a less expensive prison colony than Australia. Recognising the publicity value of official endorsements, he toured the country giving lectures and extolling the virtues of his inventions. On such occasions he proudly sported on his chest the seven medals – four gold and three silver – presented to him by various heads of state, and scientific and

Captain Manby developed the rescue mortar and other life-saving devices. (Norfolk County Council)

humane societies. Retiring from his post as the Great Yarmouth Barrack-Master in 1836, he became convinced there should be a public memorial to the success of the Manby Mortar – so he had one specially made. No coastal town would agree to accept it, so he sited it in his own garden!

Two years before his death in 1854 Manby privately produced his memoirs, from which it becomes evident that the Royal Navy never did purchase his fire extinguishers. Undeterred by their lack of interest, Manby claimed to have developed a domestic version (which could also be used in boats), and even at that time in his life was still forlornly hoping it might yet be of public service.

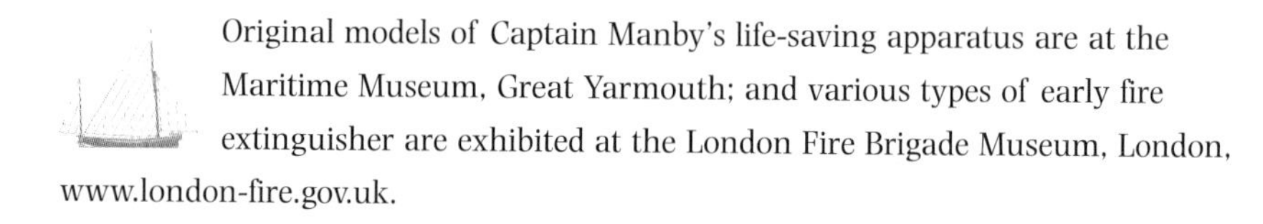

Original models of Captain Manby's life-saving apparatus are at the Maritime Museum, Great Yarmouth; and various types of early fire extinguisher are exhibited at the London Fire Brigade Museum, London, www.london-fire.gov.uk.

Peter Halkett's Inflatable Cloak-boat

In 1823 a Glaswegian called Charles Macintosh, who had already made a fortune by inventing bleaching powder, patented a revolutionary method of rendering two canvas fabrics waterproof by uniting them with a solution of naphtha and India-rubber. By April 1824 word of the wondrous material had spread to the explorer John Franklin RN, who, for his 1825–7 expedition to the Arctic, requested a number of mackintosh air-beds and pillows, and a large quantity of mackintosh sheeting for covering wooden-frame boats.

The idea of using flotation bags to convey goods over water was far from unique: Assyrian soldiers had reputedly been ordered to use them to cross a river in 880 BC, and indeed as late as 1957 inflated goatskins were still being used for raft-making on the Yellow River in northern Tibet. In 1839, for the purposes of improving Britain's military bridge-building equipment, the Duke of Wellington himself tested inflatable pontoons made of mackintosh-type fabric. However, the first person to make an inflatable *boat* was Lieutenant Peter Alexander Halkett RN. Born in about 1820, he entered the Navy in 1835, passed for Lieutenant in 1840, and served in the Mediterranean and Far East. He lived on the brow of Richmond Hill in a distinguished house called The Wick (owned many years later by the Rolling Stones guitarist Ronnie Wood). The house actually belonged to his father John Wedderburn Halkett, a director of the Hudson's Bay Company.

The prototype rubber boat that Peter Halkett began to develop there in 1842 was oval in shape, some 7ft 1in long and 3ft 5in broad, with an inboard space 3ft 5in by 1ft 4in. He called it the 'cloak-boat' because it was ingeniously glued within the lining of a 9ft-wide semicircular mackintosh cloak! When afloat, the edges of the cloak folded inwards to enable access to pockets containing a rubber air pump (the craft could reputedly be inflated in 30 seconds), a puncture repair kit, inflatable paddles and even a walking stick which contained an umbrella sail! The first experimental trip made by Halkett in the cloak-boat was along the 11-mile stretch of the Thames from Kew to Westminster, during the course of which he was nearly run down by several metropolitan paddle-steamers. By 1848 the published prospectus of the boat stated that he had subsequently paddled it at Brighton, Plymouth, Spitzbergen and Brighton again.

Halkett's golden opportunity to really show what his invention could do came in November 1844 when he was asked to demonstrate it to the Navy's Experimental Squadron in the Bay of Biscay. Conditions were choppy, but by means of the paddles and the eccentric umbrella sail he succeeded in departing from

Peter Halkett, the naval officer from Richmond-upon-Thames who invented the cloak-boat in 1842. (Portrait by George Engleheart)

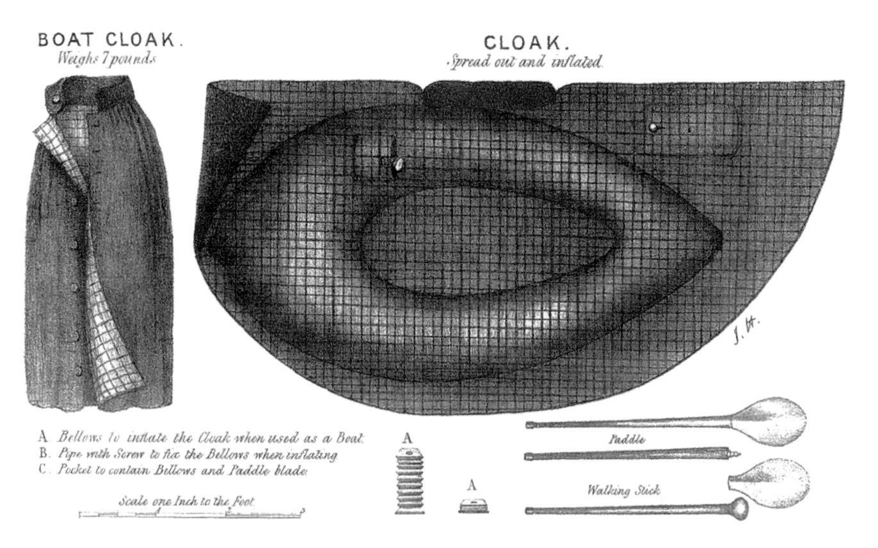

Used on Arctic expeditions, the versatile cloak-boat consisted of an inflatable ring within a cloak (complete with air-pump and paddle). (Halkett, *Boat-Cloak or Cloak-Boat,* 1848)

HMS *Caledonia* and arriving at HMS *St Vincent* safe, sound and dry. Sir David Milne, commanding the squadron, wrote to the Admiralty about Halkett's boat towards the end of January 1845, and their Lordships duly ordered a boat of that type and one of Halkett's larger boats to be made and tested at Portsmouth. It all seemed to be going well. But, Halkett's presentation was becoming flawed by confusion about what he was really trying to achieve. Was the purpose of his cloak-boat to save lives in the event of a shipwreck? Was it a glorified rainproof cloak for sartorially minded officers? Was it only intended to be used on expeditions of discovery? The options were narrowed by officialdom in May 1845 when the Second Secretary of the Admiralty, John Barrow, wrote to Halkett describing the boat as 'extremely clever and ingenious' and suggesting that 'it might be useful in Exploring and Surveying

It even had an umbrella sail! (Halkett, *Boat-Cloak or Cloak-Boat*, 1848)

Expeditions'. However the Lords of the Admiralty apparently did 'not consider that it would be made applicable for general purposes in the Naval Service'.

Halkett's 1848 publicity booklet illustrated the cloak-boat's portability, emphasising that the entire craft could be completely folded and packed in a common knapsack. So perhaps his main intention all along had been that it should be used by survey parties as a means of crossing rivers. Interestingly, the earliest person to place an order for the Halkett boat was Sir John Franklin (although it is not known for certain if he took it with him in 1845 on his ill-fated expedition in search of the Northwest Passage). Several other British explorers also used the cloak-boat in the Arctic, the most enthusiastic of whom was Dr John Rae, who found it to be enormously useful for setting and examining fishing nets. Indeed, he even named a part of the coast on the Melville Peninsula after Halkett. Parties sent in search of Franklin were equipped with an improved version of the cloak-boat (specially protected with cork fenders), and in 1854 it was Rae himself who discovered the first relics of the lost expedition. Dr Rae took his own cloak-boat home to the Orkneys, where in 1954 it was discovered in the loft of a Kirkwall timber yard and then deposited in a museum in Stromness.

Few significant steps were taken to develop Halkett's invention until 1937 when an engineer called Pierre Debroutelle, working for the French military balloon company Zodiac, designed a horseshoe-shaped inflatable boat. Known as the 'A'-type, it was rowed by Debroutelle on the Seine in 1940 and tested with a borrowed outboard motor, then in 1943 its unique wooden transom was patented. Initially almost the only postwar customers of the Zodiac were French commandos. That all changed in 1952 when Dr Alain Bombard purchased a standard 15ft Zodiac, which he named *l'Hérétique*. Equipped with just a 27 sq. ft lugsail and a leeboard, he succeeded in sailing it from the Canaries across the Atlantic to a beach in Barbados in just sixty-four days. By making the voyage without supplies of food or fresh water, relying only on rainwater and what fish he could catch, Bombard was endeavouring to prove that the right survival methods could greatly increase the chances of existence for shipwrecked people. In France and around the world Bombard's crossing had an enormous impact, and his book *Shipwrecked at Will* became a bestseller. Unintentionally he gave inflatable boats an adventurous image, which helped sales of them in the leisure market.

The means by which inflatables could be made more seaworthy, as well as faster, was devised by the founding headmaster of Atlantic College in South Glamorgan, Admiral Desmond Hoare. There in 1963 Hoare made the first rigid inflatable, by ingeniously gluing a plywood floor to an inflatable dinghy. In the 1969 'Round Britain Power Boat Race' another such boat, *Psychedelic Surfer*, became the lowest-powered boat to complete the course. The V-shaped rigid lower hull Hoare developed with Avon Inflatables, after trials facilitated the growth of the RNLI's quickly expanding fleet of inshore rescue craft. By then the British RFD Company, founded by an entrepreneur called Reginald Foster Dagnall, had succeeded in 1932 in devising a rubber liferaft, and in partnership with Zodiac was able to profit from a French government decree of 1955 requiring *all* merchant ships to be equipped with inflatable liferafts.

Yet Peter Halkett had been making plans for inflatable liferafts a century earlier. In a footnote to his 1848 prospectus he proposed the construction of a 30ft by 16ft inflatable raft which could take 30–40 persons; when uninflated and rolled up, it occupied a space no larger than that a spare sail might occupy. He went on: 'Ashore it might be easily carried on horseback – paddles and all – to any parts of the coast where its services might happen to be called for.' The loss of the Franklin expedition might have occasioned Halkett to advocate such an approach, or perhaps it caused him to become disillusioned with inflatable boats. His experiments with them evidently did his naval career no harm because in the late 1840s he was promoted to captain. Thereafter, until his death in Torquay in 1865, he turned his attention to other inventions, such as a system for ploughing fields with steam tractors, an improvised inkstand, and an apparatus for raising ships that had sunk.

The earliest Halkett inflatable known to be still in existence is Dr John Rae's cloak-boat at the Stromness Natural History Museum, Orkney I.; Alain Bombard's dinghy *l'Hérétique* is at the Conservatoire International de Plaisance, Bordeaux, www.bordeaux-city.com.

Edward Berthon's Folding Dinghy

The ingenious stowable dinghy much prized by Victorian yachtsmen as a tender had originally been designed as a life-saving boat. Its creation was prompted by a tragic accident on 29 June 1849, when the SS *Orion*, one of the earliest iron-built passenger steamers, struck a submerged rock and sank off Portpatrick on the Scottish coast. Fatally the liner had carried only two lifeboats which the crew seized for themselves, leaving nearly all her 150 passengers to drown. Almost the only survivor was a Stretford clergyman who managed to cling on to some floating luggage. Adrift and helpless, he figured that such heavy loss of life might have been avoided had *Orion* carried space-saving emergency boats. Rescued from the sea by local fishermen, he wrote to the vicar of a parish near Portsmouth who was reputed to be a keen amateur inventor of maritime equipment, and asked him: 'Can you, with all your nautical genius, think of a means by which boats enough for all on board can be stowed on a passenger steamer?'

Within hours of the letter's delivery at Fareham the Revd Edward Berthon had figured out just what was needed: collapsible boats formed like the segments of an orange, with curved longitudinal timbers, jointed at the ends, and covered with waterproofed canvas in place of planking. Berthon calculated that four of his 'Collapsibles' would occupy the deck space of one conventional lifeboat.

The invention was a remarkable achievement (even for this astonishingly prolific inventor who during his lifetime would achieve twenty-five patents to his name). But frustratingly, Berthon had hitherto always been commercially unlucky, forever just missing out on getting his ideas accepted. As long ago as 1835, as a young theology student, he reputedly became the first Englishman to invent a working two-bladed propeller, but the Admiralty had

Berthon's folding dinghies, though originally intended to be lifeboats, were much prized by late-Victorian leisure boaters as yacht tenders. (Dixon Kemp, *A Manual of Yacht and Boat Sailing*, 1900)

gruffly dismissed his system, saying it 'never would and never could propel a ship'. Next he invented the Berthon Log, a remarkably futuristic hydraulic mechanism for recording a vessel's speed. But the Admiralty rejected that too, as they did the clinometer, an instrument for measuring a ship's heel and pitch.

Several 30ft prototypes of the collapsible boat – which could carry ninety persons – were built in the vicarage gardens and were successfully trialled at sea, even being awarded a medal at the 1851 Great Exhibition. But it won no orders. Merchant shipowners were not interested. At Cowes Berthon was commanded to demonstrate the boat to Queen Victoria, who was so impressed that Prince Albert advised the Admiralty to give it a try. Subsequently a technical evaluation was done on the collapsible by the Portsmouth dockyard, with excellent results. Having spent years pressing other inventions on the Admiralty, Berthon knew he had accumulated high-ranking enemies, but even so he was shocked in 1858 to be informed that their Lordships had emphatically rejected it. To the good-natured and jovial Berthon, a Huguenot aristocrat who had devoted his entire working life to helping others, this was a crushing blow. Bitterly disappointed, he burnt all the prototype boats and resigned from Fareham, determined to get as far away as possible from the sea and the Royal Navy.

Berthon busied himself restoring the abbey church at Romsey where he had become the new vicar. The folding boat might then have been forgotten altogether were it not for the encouragement of the chief parishioner at Broadlands there, Lord Palmerston, and the intervention of the maritime safety campaigner Samuel Plimsoll, who in 1873 persuaded Berthon to look at it again. He duly applied himself to perfecting the collapsible boat. He added another layer of canvas to make the boats double-skinned, and fussed over details such as the special mixture of paint used to waterproof the canvas. This time he met with complete success! Before the end of the year Berthon had received orders from the Admiralty for £15,000-worth of new boats, the first of which were sent to General Gordon at Khartoum. Soon several other companies were hastily producing imitation Berthons. The popularity of the folding boat then grew so quickly that by 1877 Berthon had established a substantial boatbuilding concern, the Berthon Boatyard, at Lymington. Amazingly, more than a quarter of a century had passed since the initial invention!

Edward Berthon's zest for invention was rekindled by this success, and he went on to produce for the Army a portable bridge pontoon variant of the lifeboat, and even in his eighties made another invention – the folding military bandstand. Berthon's son took over the direction of the boatbuilding business in 1892, widening their product range to include dinghy-sized versions of the folding lifeboat. Proving ideal for use in the Royal Navy's small destroyers, and even in early submarines, the dinghies soon became available to amateur yachtsmen who marvelled that they could be squeezed through

Revd Edward Berthon was a prolific inventor of maritime equipment. (Berthon, *A Retrospect of Eight Decades*, 1899)

The Self-draining Cockpit

Now regarded as a crucial safety measure, the self-draining cockpit was invented in 1889 as a competitive device by an American, Paul Butler, who at the time was the world's leading authority on the building and sailing of racing canoes. That year Butler produced a sailing canoe called *Fly*, which, in addition to the self-draining cockpit, had a few other radical new inventions such as a sliding seat, which enabled the helmsman to sit outside the hull and still control the sails, and clutch cleats to secure reefing lines.

hatches just 14in across. Then in 1912 disaster struck. Four of the *Titanic*'s much-criticised lifeboats were of the Englehart folding type. It made no difference whether they were made by Berthon's or by other companies, the loss of confidence in them was such that the era of the Berthon collapsible lifeboat came to an abrupt end.

Nevertheless, the Berthon's stowing qualities had made it popular among pioneering cruising yachtsmen. Claud Worth, in his 1910 book *Yacht Cruising*, claimed to have had 'Berthon boats of 7, 8 and 9 feet length. The 8-foot is the most generally useful. It is easily got on board single-handed. It will carry two people, or three at a pinch. With average skill it is quite good enough for working a 40-lb. kedge.' Writing in 1895, Edward Knight commented: 'In my opinion one cannot do better than carry a medium-sized Berthon collapsible. A Berthon occupies very little room, and is so easily dropped into the water and hoisted on board again that she is sure to be used on many occasions.'

Historic Berthon dinghies are on display at Brodsworth Hall, South Yorkshire, www.sheffnet.net; and Unsted Boat Haven, Haroldswick, www.unst.shetland.co.uk. See also the Berthon Boat Company, Lymington, www.berthongroup.co.uk.

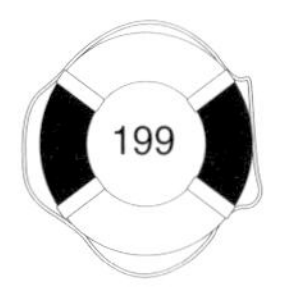

John Ward's Lifejacket and the First Survival Suits

What was to become the first standard lifebelt was designed in 1854 by Captain John Ross Ward RN, who at time was the Inspector of Lifeboats for the National Lifeboat Institution. Since the late eighteenth century various inventors had experimented with devices whose purpose was to save lives at sea. One of the earliest devices for supporting a man in the water was a cork lifejacket designed in 1788 by a Paris boatbuilder who jumped into the Seine to demonstrate the effectiveness of his jacket, which was simply a bundle of cork strips tied together with string. Then in 1804 a choleric clergyman called William Mallison produced a more advanced version, which consisted of slabs of cork worn front and back and joined by tapes. Mallison spent a fortune promoting his design, but the jacket proved unstable in the water and had the major disadvantage of requiring a range of sizes to fit.

In 1852 Captain Ward commenced a thorough survey of the many different types of lifejacket then available. Dismissing them all, especially those with air chambers, which he feared could easily rip, he decided to design his own version. Introduced in 1856, it was a compact design, consisting of strips of cork stuck to a canvas frame. Lifeboatmen were expected to don it every time they went afloat. But according to an article on this subject by Martyn Clemans in 1965, many lifeboatmen refused to do so, regarding such practices as cowardly. Cork lifejackets had several disadvantages. They were liable to injure the wearer if they jumped into the sea from any height, and the movement of the waves caused them to chafe the skin on neck and chin. Nevertheless Captain Ward's lifejacket was later introduced into the US forces, and was a basic safety item until 1906 when the first Kapok type of jacket was introduced.

Another early life-saving device was the lifebuoy. It is usually thought that the now-ubiquitous horseshoe lifebuoy on yachts is quite a modern

The cork lifejacket devised by John Ward in 1854. (Lewis, *The Life-Boat and its Work*, 1874)

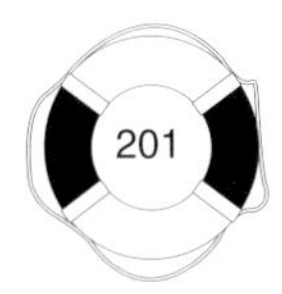

design, but in fact it all began two centuries ago with Alfred Kynaston's 'Life-Buoy', a horseshoe-shaped arrangement of cork blocks set in a frame of canes and all covered in canvas smeared with tar. A flagstaff fitted to it provided a handhold for the victim to cling on by. Eventually the Board of Trade accepted as standard a circular Kisbie cork lifebuoy.

Survival suits are also nothing new. John Bentley's 'nautilus' suit of 1797 consisted of a three-chambered copper belt worn around the waist, and fixed to the body with straps. It had a covering of baize and consisted of swimming gloves and frog-feet of leather supported by wooden struts. In 1831 a New York inventor called Henry Bateman had the idea that every passenger on a ship should have their own survival capsule – complete with cigars and newspapers! He nearly died when experimenting with just such a survival suit at the 1831 Naval Exhibition, when the lid of the capsule he was testing on the Hudson River failed to open. In terms of foulweather gear, sou'wester hats made from tarpaulin (canvas covered with tar) went on sale in 1841, and by 1894 a former Scandinavian sea captain called Helly Juell Hansen, who since 1877 had supplied utilitarian waterproofs for fishermen, was providing stylish oilskins for yachtsmen – which he made from unbleached cloth impregnated with linseed oil!

The Royal National Lifeboat Institution website, www.rnli.org.uk.

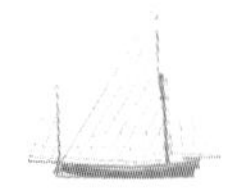

Reflective Tape

The reflective patches that can so readily be seen in a searchlight beam on lifejackets, liferafts and rescue beacons are called Scotchlite. The retro-reflective sheet from which they are cut was invented in 1938 by Philip Palmquist, an engineer at 3M Sellotape, who had been endeavouring to find a means of increasing the efficiency of road signs.

Martha Coston's Signalling Flare

The memorable scene in the movie *From Russia with Love*, when James Bond, being chased in a Fairey Huntress powerboat along the Dalmatian coast, pulls out a signal flare and sets alight his SPECTRE pursuers, would not have been possible without the use of a flare pistol. The one used in the film was the ubiquitous Very pistol, named after Lieutenant Edward Wilson Very of the US Navy, who roguishly claimed to have invented the device in 1877 – although the real originator of the distress flare was an American widow called Martha Coston.

Her husband, Benjamin Franklin Coston, was French by descent but had been born in Philadelphia. A brilliant pyrotechnist, he had landed a job at the Washington Navy Yard where he created and supervised a pyrotechnic laboratory. There he built improved Hales rockets for use in the Mexican War, and is also said to have developed a prototype submarine. Something of a would-be entrepreneur, at the age of sixteen he had founded a company to develop his inventions. Significantly he invented a cannon percussion primer, which, although he was a Navy employee, he developed as his own property. Their derisory offer of compensation for the invention prompted him in 1847 to resign. Soon selected to run the Boston Gas Company, he invented the first ever portable gas apparatus, and then an advanced form of gaslight (indeed the first gas burnt in the city was said to have been made in Coston's own residence). The inhalation of various chemicals and gases in the Navy Yard had taken a toll on his health and in November 1848, aged only twenty-four, he died unexpectedly of pneumonia.

Martha's life with her husband had set a frantic pace. Originally from Baltimore, she was said to have eloped with the promising engineer Benjamin when she was only fourteen. His death left her a widow at twenty-one with four young children to support. The bankruptcy of Benjamin's company soon meant she was virtually penniless. But a chance prospect of salvation appeared one afternoon when she opened a wooden trunk to sort through her late husband's papers. There, among various bundles of correspondence, she discovered a sealed notebook containing some rough charts and outlines of an unfinished invention, which Benjamin had evidently considered too costly for his company to develop.

The plans were for a night signalling flare, and seeing them for the first time jogged Martha's memory: Benjamin had prepared some prototypes at the Washington Navy Yard. The device was intended to be a hand-held launcher, with a simple spring firing hammer, which would fire from an attached tube combinations of red, white and blue flares. Martha desperately needed the signal flares to succeed to enable her to support her children, and yet she had no experience of business, knew nothing of the process of invention, and had no scientific knowledge. When in due course she did retrieve those early flares their damaged condition confirmed her fears of the immense struggle she would face to get the Navy to help her develop the invention.

However, Martha had some High Society contacts in Washington, and she succeeded in getting a charmingly persuasive message to the Secretary of the Navy, Isaac Toucey. He arranged for some of the necessary research to be paid for from the Navy's contingency fund, although the Navy itself would not develop the system. Still smarting from the row about the percussion caps, the Navy was unenthusiastic, so Martha was quite unsurprised to be informed time after time that the flares had failed in trials. Although she had succeeded in getting a bright white and a vivid red, she could not obtain a strong enough blue – and it was essential that the flares were of equal intensity in order to make out the necessary transposition for the signal figures. Martha had set her heart on the third colour being blue because, with red and white, it was one of the national colours.

For almost a decade she persevered with her experiments until the answer occurred to her in 1858 while watching a massive firework display celebrating the successful laying of the first transatlantic telegraph cable. Realising that firework technology could provide the necessary answer, she worked with a leading New York pyrotechnist to develop a strong green colour that was as brilliant as the red and the white. This enabled Martha in 1859 to patent the invention – albeit in her late husband's name – and to establish a factory to make the flares.

A Naval Board of Examiners having approved the new 'Coston night signals', that year the Navy proceeded to place a $6,000 order with her factory for Coston sets, each of which contained a dozen flares. Martha then headed for Europe. Travelling around the continent with her explosive flares in a trunk marked 'music boxes', she secured patents in seven countries. It is estimated by Coston's biographer, Denise Pilato, that during the Civil War the US Navy was supplied by the Coston factory with some 1.2 million flares for use both in ship-to-ship and ship-to-land communication. Some $120,000 should have been due to Martha for that, but eventually the government only offered her $15,000.

The flares were increasingly used as danger and distress signals, and in 1871 Coston's produced an improved device which they marketed on the strength of its safety features. Patented that year, it was fired by a new twist-grip mechanism Martha had invented herself. To promote the product she collected and distributed testimonials from newspapers, mariners and yacht clubs. The improved device ought to have established Martha, she reckoned, as an inventor. Thus she was indignant when a US Navy lieutenant called Edward Wilson Very in 1877 adapted her flares to be fired from a simple shotgun-specification pistol. Marketed worldwide (in Britain by Webley & Scott), his innovative launch mechanism soon became known as the 'Very pistol'. For the rest of her years Martha struggled to get the Very pistol known as the 'Coston pistol'. This perceived injustice encouraged her to produce her remarkable autobiography, *A Signal Success: The Work and Travels of Mrs. Martha J. Coston*, which was published eighteen years before her death in 1904.

The Coston Signal Company she had founded remained in the family and went on to produce line-throwing guns and other rescue equipment. In 1954 international agreement was finally reached that distress flares should be red. For a time the Very pistol continued to be regarded as the most thrilling means of firing them off, but in recent years it has been abandoned in favour of the simpler hand-launched device pioneered by Martha Coston.

Guglielmo Marconi's S-O-S

Since 1896, when the Italian-born Guglielmo Marconi arrived in England and successfully demonstrated on Salisbury Plain that his revolutionary wireless system could send Morse code messages, the use of wireless telegraph systems had spread rapidly. In 1899 he established wireless communication between France and England across the English Channel, and in the following year founded the Marconi Maine Communication Company.

COMMONEST CAUSES OF RNLI LIFEBOAT CALL-OUTS

1. Machinery failure
2. Stranding/grounding
3. Adverse weather conditions
4. Capsize
5. Fouled propeller

Within months the wireless telegraph (w/t) proved it could save lives when a wireless call enabled the crew of the Borkum Light Vessel to be rescued when their lightship broke her moorings in a storm. Initially distress calls were prefixed with the letters CQ, meaning 'all stations', but as the same prefix was also used for other general calls, it was not particularly distinctive. So in 1904 the Marconi Company ordered that all ships controlled by them should use CQD. It did not mean 'Come Quick Danger' or 'Come Quick Distress', as is popularly assumed, but the 'D' did mean 'Danger'. Marconi's pioneering use of such a simplified distress signal eventually set a precedent and at the 1906 Radiotelegraphic Conference it was agreed that SOS should become the international distress signal. It was chosen not because it meant 'Sink Or Swim' or 'Save Our Souls', but because the Morse letters were easy to remember.

There are various early radios at the Science Museum, London, www.sciencemuseum.org.uk.

Daniel Mowrey's Seasickness Remedy

From time immemorial the causes of seasickness or *mal-de-mer* remained a mystery. The engineer Sir Henry Bessemer wondered if it might be preventable with a more stable form of ship, so in 1875 he built a cross-Channel steamer within whose hull the passenger accommodation could pivot in order to keep it level regardless of the ship's movement. But unfortunately the experiment failed because passengers became even more sick when the artificial flooring got out of step with the motion of the sea.

Naval medical literature from Elizabethan times onwards records many instances where a ship's handling and fighting capacities were affected when many of the crew became seasick. Yet,

curiously, no sustained attempt was made by the Royal Navy to establish the causes of seasickness. In the absence of a standard means of dealing with it, quack cures – such as chicken broth – were sometimes attempted. Some travellers, according to one limerick, even tried champagne:

> There once was a young lady from Spain,
> Who washed down her meal with champagne,
> In a futile attempt
> To make her exempt
> From seeing the same food again!

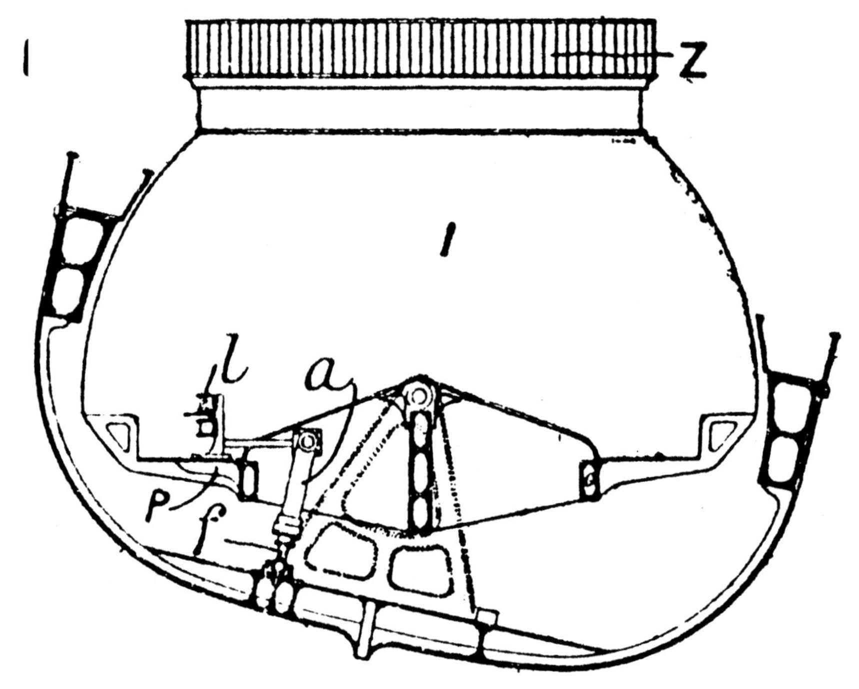

Above and facing page: In 1870 two ingenious ideas to prevent seasickness were patented by Sir Henry Bessemer, an engineer with ambitions to equip Channel ferries with self-balancing saloons. (Patent Office, 1870)

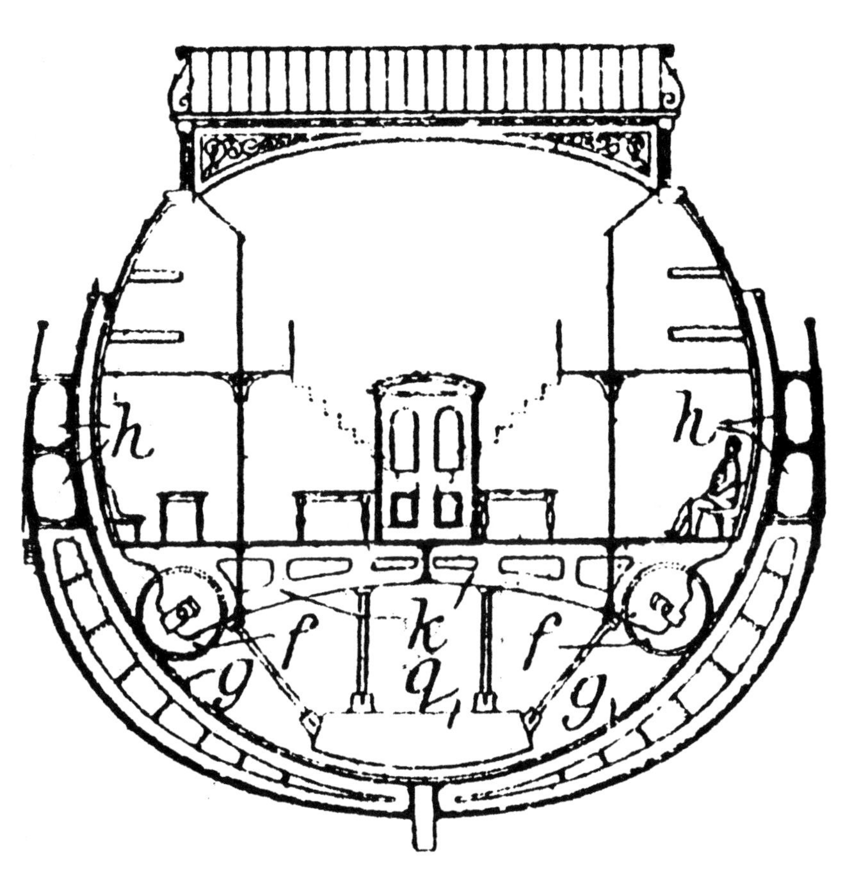

Even the seasoned yacht skipper Captain John Illingworth RN was perplexed about how best to deal with seasickness. In his definitive 1958 book *Offshore*, he acknowledged seasickness as 'the racing skipper's Public Enemy Number One', but could only offer a few somewhat unscientific suggestions, including barley sugar, Andrew's Liver Salts, Milk of Magnesia, a travel tablet, cod liver oil pills, and the avoidance of fats, red meat and alcohol.

After the Second World War it was established that seasickness was a branch of a wider malady called 'Motion Sickness', which

included car sickness, air sickness and (a few years later) space sickness. All induced almost identical symptoms. The four variable causal factors in motion sickness were environmental, physical, emotional and social. The first reliable motion sickness pill was an antihistamine called Dramamine, which was developed at the John Hopkins University in Maryland and marketed by Polaroid. Various other preventative seasickness drugs (such as Dimenhydrinate) were also being developed, but these tended to cause drowsiness – which could itself be a danger at sea.

These scientific certainties were shaken in March 1982 by an article in *The Lancet* showing the favourable results achieved among the suffers of motion sickness by the use of ginger. The article was principally authored by an American doctor, Daniel Mowrey of Utah. Mowrey had discovered that as long ago as 1597 the famous London herbalist John Gerald had recommended ginger for its mitigatory effects on gastrointestinal distress. Indeed, the explorer Martin Frobisher had taken ginger in his medical supplies on an expedition to discover the Northwest Passage in 1576. Trials were carried out by Mowrey in which thirty-six Utah students were each placed in a tilting rotating chair to induce motion sickness. Instead of a fluid extract of rhizome of ginger, the candidates received ginger root in powdered form. The test found that ginger in a powdered form was superior to dimenhydrinate in preventing the gastrointestinal symptoms of motion sickness. Mowrey concluded that the aromatic and carminative properties of ginger ameliorated the effects of motion sickness in the gastrointestinal tract, and also absorbed toxins and acids. Tests then done by the Australian and Danish navies found that in heavy seas ginger could reduce vomiting and cold sweating. It all seemed to show that Mowrey's use of ginger was effective, and that it could provide a useful remedy for seasickness without the disadvantageous side-effects of high-tech inventions.

Glossary

Abreast	Directly at right angles to fore and aft line
Awash	Level with the water surface
Bare poles	Having no sails set
Beam	Breadth of vessel at widest point
Boom	Spar used to extend foot of a sail
Bowline	Versatile knot, and a line keeping square-sail to windward
Centreboard	Pivoted retractable keel
Close-hauled	Sailing as close to the wind as possible
Cutter	Single-masted vessel with two headsails
Dagger-board	Retractable board through keel to reduce leeway
Fore and aft rig	All canvas set on fore and aft line
Forefoot	Foremost end of the keel
Foresail	A cutter's foremost jib
Freeboard	Distance between waterline and gunwhale
Gooseneck	Hinged joint attaching heel of boom to mast
Hank	Clip or hoop for attaching luff of sail
Heave to	Virtually stop
Heel, to	To list under sail
Jib sheet	Rope by which jib is trimmed
Ketch	Two-masted sailing vessel, the mizzen mounted forward of rudder post
Lee side	Opposite side from wind direction
Leeboard	Board lowered on the lee side
Leeward	Downwind

Luff, to	To bring boat's head closer to the wind
Lugsail	Four-sided sail, its head bent to a yard, slung to leeward of the mast
Mainsail	Large sail hoisted on a mainmast
Mainsheet	Rope by which mainsail is trimmed
Mizzen	Sail hoisted on the mizzen-mast
Pitch	Fore and aft rocking motion
Planing	Hull skimming over surface of water
Port	Left-hand side of a vessel
Reach	Sail across the wind
Rigging screw	Bottle-screw or turnbuckle for tautening stays
Run, to	To sail with the wind aft
Running rigging	The movable parts of a vessel's rigging
Schooner	Vessel with two or more masts, the mainmast to the aft
Sharpie	American oyster boat
Shroud	Wire rope supporting a mast from the side
Spreaders	Struts fitted to the mast to spread the shrouds
Standing rigging	Parts of rig set permanently
Starboard	Right-hand side of a vessel
Swing the lead	Heave the lead by under-arm swinging
Tack	Alternating course to progress close-hauled to windward
Weather side	Side on which the wind blows

Further Reading

The research for this book was chiefly done at the National Maritime Museum, Greenwich; the Science Museum Library, South Kensington; the British Library; the Marine Society; and the Cruising Association, and I am thankful to the librarians of those establishments for their assistance and cooperation.

GENERAL

Baader, Juan, *The Sailing Yacht* (Adlard Coles, 1965)

Bond, Bob, *The Handbook of Sailing* (Pelham, 1996)

Burgess, F.H., *A Dictionary of Sailing* (Penguin, 1961)

Charnock, John, *An History of Marine Architecture* (Faulder, 1802)

Clark, Arthur, *The History of Yachting, 1600–1815* (Putnam, 1904)

Classic Boat, 'Top 150 Boats' (December 2000)

Coffin, Roland Folger, *The History of American Yachting* (1887)

Cooke, Francis, *Hints, Tips and Gadgets* (Edward Arnold, 1939)

Cooke, Francis, *The Corinthian Yachtsman's Handbook* (Edward Arnold, 1913)

de Vries, Leonard, *Victorian Inventions* (John Murray, 1971)

Dear, Ian, *The Champagne Mumm Book of Ocean Racing* (Severn House, 1985)

Desmond, Kevin, *The Harwin Chronology of Invention, Innovations and Discoveries* (Constable, 1987)

Eastland, Jonathan, *Great Yachts and their Designers* (Rizzoli, 1980)

Folkard, Henry, *Sailing Boats from Around the World* (Chapman and Hall, 1906)

Folkard, Henry, *The Sailing Boat* (Edward Stanford, 1901)

Giorgetti, Franco, *History and Evolution of Sailing Yachts* (White Star, 2000)

Gougeon, Meade and Knoy, Ty, *The Evolution of Modern Sailboat Design* (Winchester, 1973)

Hartman, Tom, *The Guinness Book of Ships and Shipping Facts & Feats* (Guinness Superlatives, 1983)

Heaton, Peter, *Yachting: A History* (Batsford, 1955)

Henderson, Richard, *Singlehanded Sailing* (International Marine, 1992)

Herreshoff, Francis, *An Introduction to Yachting* (Sheridan House, 1967)

Herreshoff, Halsey (ed.), *The Sailor's Handbook* (Marshall, 1999)

Johnson, Peter, *The Encyclopedia of Yachting* (Dorling Kindersley, 1989)

Johnson, Peter, *The Guinness Book of Yachting Facts and Feats* (Guinness Superlatives, 1975)

Johnson, Peter, *The RYA Book of World Sailing Records* (Adlard Coles, 2002)

Kemp, Dixon, *A Manual of Yacht and Boat Sailing* (Cox, 1891)

Kemp, Peter, *The Oxford Companion to Ships and the Sea* (OUP, 1988)

Knox-Johnston, Robin, *History of Yachting* (Phaidon, 1990)

Leather, John, 'The Forgotten Designers', *Classic Boat* (January 1977), pp. 36–9

Martin-Raget, Gilles, *Legendary Yachts* (Abbeville Press, 2002)

Neison, Adrian, Kemp, Dixon, and Davies, Christopher, *Practical Boat Building and Sailing* (Upcott Gill, 1900)

Phillips-Birt, Douglas, *The History of Yachting* (Elm Tree Books, 1974)

Schult, Joachim, *Curious Boating Inventions* (Paul Elek, 1974)

Schult, Joachim, *Curious Yachting Inventions* (Paul Elek, 1974)

Seidman, David, *Sailing* (Adlard Coles Nautical, 2001)

Stephens, William, *American Yachting* (Macmillan, 1904)

Thompson, Elaine, *100 Years of Yachting World* (Yachting World, 1994)

Useful Websites

American Maritime Museums	www.maritimemuseums.net
American Sailing Associations	www.american-sailing.com
Biography	www.biography.com
Classic Boat	www.classicboat.co.uk
Cruising Association	www.cruising.org.uk
Cruising Club of America, The	www.cruisingclub.org
European Classic Yacht Union	www.ecyu.org
International Sailing Federation	www.sailing.org

Inventors	www.inventors.about.com
Musée National de la Marine, Paris	www.musee-marine.fr
Museum of Yachting	www.moy.org
Mystic Seaport Museum	www.mysticseaport.org
National Historic Ships Committee	www.nhsc.org.uk
National Maritime Museum, Cornwall	www.nmmc.co.uk
National Maritime Museum	www.nmm.ac.uk
Newport News Museum	www.mariner.org
Patent Office	www.patent.gov.uk
Pathe News	www.britishpathe.com
Practical Boat Owner	www.pbo.co.uk
Royal Yachting Association	www.rya.org.uk
World Sailing Speed Record Council	www.sailspeedrecords.com
Yachting Monthly	www.yachtingmonthly.com
Yachting World	www.yachting-world.com
Yacht Racing Links	www.semaphore.co.uk

CHAPTER 1. SAILING PIONEERS

King Charles II, Yacht Racer

Carr, Frank, *The Yachtsman's England* (Seeley Service, 1937)
Clark, Arthur, *The History of Yachting, 1600–1815* (Putnam, 1904)
Crabtree, Reginald, *Royal Yachts of Europe* (David & Charles, 1975)
McGowan, A.P., *Royal Yachts* (National Maritime Museum, 1977)
Ratsey, Ernest, and de Fontaine, Wade, *Yacht Sails: Their Care and Handling* (Norton, 1948)
Sottaa, Jules, 'Early Yachts', *Mariner's Mirror* (1920), vol. 6, pp. 115–19
't Hooft, C., 'The First English Yachts', *Mariner's Mirror* (1919), vol. 5, pp. 108–23

John Cox Stevens and the New York Yacht Club

Clark, Arthur, *The History of Yachting, 1600–1815* (Putnam, 1904)
Folkard, Henry, *The Sailing Boat* (Edward Stanford, 1901)
Giorgetti, Franco, *History and Evolution of Sailing Yachts* (White Star, 2000)
Guest, Montague and Boulton, Williams, *The Royal Yacht Squadron* (John Murray, 1902)

Heaton, Peter, *Yachting: A History* (Batsford, 1955)
Johnson, Peter, *The Encyclopaedia of Yachting* (Dorling Kindersley, 1989), pp. 22–3
Kemp, Peter, *The Oxford Companion to Ships and the Sea* (OUP, 1988)
Leather, John, 'The Rise of the America', *Classic Boat* (July 2001), pp. 38–43
Phillips-Birt, Douglas, *The History of Yachting* (Elm Tree Books, 1974), pp. 38–9
Ratsey, Ernest, and de Fontaine, Wade, *Yacht Sails: Their Care and Handling* (Norton, 1948)
Rousmaniere, John, *The Low Black Schooner* (Mystic Seaport Museum, 1987)
Stephens, William, *American Yachting* (Macmillan, 1904), pp. 8–29

Richard McMullen's Corinthian Yachting

Cooke, Francis, *Cruising Hints* (Yachtsman, 1904)
Cooke, Francis, *The Corinthian Yachtsman's Handbook* (Edward Arnold, 1913)
Guest, Montague and Boulton, Williams, *The Royal Yacht Squadron* (John Murray, 1902), pp. 76–7
Heaton, Peter, *Yachting: A History* (Batsford, 1955)
McMullen, Richard, *Down Channel* (Cox, 1893)
Ratsey, Ernest, and de Fontaine, Wade, *Yacht Sails: Their Care and Handling* (Norton, 1948)

James Gordon Bennett's Transatlantic Race

Foster, Q. Hooper, 'James Gordon Bennett Leads Yachting into the Transatlantic Scene', *Sea History* (Summer 1999), pp. 23–7
Heaton, Peter, *Yachting: A History* (Batsford, 1955)
Rayner, Denys and Wykes, Alan, *The Great Yacht Race* (Davies, 1966)
Rousmaniere, John, *The Golden Pastime* (Nautical Quarterly, 1986)
Seitz, Dom C., *The James Gordon Bennetts* (Bobbs-Merrill, 1928)
Stephens, William, *American Yachting* (Macmillan, 1904)

John MacGregor, Boating Celebrity

Dictionary of National Biography (OUP, 1921–2), vol. xii, pp. 541–2
Hodder, Edwin, *John MacGregor* (Hodder, 1891)
MacGregor, John, *A Thousand Miles in the 'Rob Roy' Canoe* (Sampson Low, 1883)
MacGregor, John, *The Voyage Alone in the Yawl 'Rob Roy'* (Grafton, 1987)

Joshua Slocum's Circumnavigation

Brassey, Annie, *A Voyage in the 'Sunbeam'* (Longman, 1878)

Cunliffe, Tom, 'Slocum's Voyage', *Yachting World* (April 1995), pp. 44–9

Henderson, Richard, *Singlehanded Sailing* (International Marine, 1992)

Nicolson, Ian, 'The *Spray* Legend: Fact or Fiction', *Yachting Monthly* (April 1995), pp. 38–41

Robert-Goodson, Bruce, *Spray: The Ultimate Cruising Boat* (Adlard Coles, 1995)

Slocum, Joshua, *Sailing Alone Around the World*, Introduction by Thomas Philbrick (Penguin, 1999)

Slocum, Victor, *Capt. Joshua Slocum* (Adlard Coles, 2002)

Teller, Walter Maques (ed.), *The Voyages of Joshua Slocum* (Adlard Coles, 1995)

Eric Tabarly, the Yachting Populariser

Boyd, James, 'Offshore Record Breakers', *Yachting World* (May 2002), pp. 48–53

Hutchinson, Marcus, 'Growth of a Legend', *Yachting World* (May 1994), pp. 54–7

Martin-Raget, Gilles, *Legendary Yachts* (Abbeville Press, 2002)

Pace, Franco, *William Fife* (Adlard Coles, 1998)

The Times, Obituary, 16 June 1998

Yachting Monthly, 'The Black Bird Flies' (August 1996), pp. 40–1

Yachting World, 'On the Wind' (August 1998), pp. 10–16

Yachting World, 'Milestones in Yachting' (January 2000), p. 72

CHAPTER 2. NAVIGATION AND PILOTAGE

Greenvile Collins's Coastal Charts

Admiralty, *Manual of Seamanship* (HMSO, 1932), vol. ii

De Beer, Esmond (ed.), *The Diary of John Evelyn* (OUP, 1955), vol. iv

Dyer, Florence, 'Journal of Grenvill Collins', *Mariner's Mirror* (1928), vol. 28, pp. 197–219

Kemp, Peter, *The Oxford Companion to Ships and the Sea* (OUP, 1988), p. 158

Richie, G.S., *The Admiralty Chart* (Pentland, 1995)

Robinson, A., *Marine Cartography in Britain* (Leicester University Press, 1962)

Verner, Codie, 'Captain Collins' *Coasting Pilot*', *Map Collectors' Circle* (1969), No. 58, pp. 5–47

Wallis, Helen, 'The Role of the "Amateur" Hydrographer in the Eighteenth Century', *Mariner's Mirror* (1964), vol. 50, pp. 75–7

John Campbell's Lunar Sextant

Bennett, J.A., *The Divided Circle* (Phaidon Christie's, 1987)

Cardoza, Rod, *Evolution of the Sextant* (West Sea Co., 2004)

Maskelyne, Nevil, *The Nautical Almanac* (Commissioners of Longitude, 1768)

May, W.E., 'Hadley's Quadrant', *Mariner's Mirror* (1954), vol. 41, no. 1, pp. 80–1

Rosser, William, *A Self Instructor in Navigation* (London, 1863)

Warner, Deborah Jean, 'American Octants and Sextants: The Early Years', *Rittenhouse* (1989), vol. 3, no. 3, pp. 86–9

Warner, Deborah Jean, 'John Bird and the Origin of the Sextant', *Rittenhouse* (January 1988), pp. 1–11

Williams, J.E.D., *From Sails to Satellites* (OUP, 1992)

Edward Massey's Distance Log

Desmond, Kevin, *The Guinness Book of Motorboating Facts and Feats* (Guinness Superlatives, 1979)

Falconer, William, *An Universal Dictionary of the Marine* (Cadell, 1780)

Kemp, Peter, *The Oxford Companion to Ships and the Sea* (OUP, 1988), pp. 492–3

Neison, Adrian, Kemp, Dixon and Davies, Christopher, *Practical Boat Building and Sailing* (Upcott Gill, 1900)

Sharpe, A.J., *Distance Run: A History of the Patent Ship-Log* (Brassbounders, 1999)

Smyth, Charles, *Description of New or Improved Instruments for Navigation and Astronomy* (University of Edinburgh Press, 1955)

Francis Crow and the Ship's Compass

Bennett, J. A., *The Divided Circle* (Phaidon Christie's, 1987)

Bristow, H.R., 'The Crows of Kent', *Bulletin of the Scientific Instrument Society* (1998), no. 58, pp. 22–5

Carr, Frank, *The Yacht Master's Guide* (Davies, 1940)
Desmond, Kevin, *The Guinness Book of Motorboating Facts and Feats* (Guinness Superlatives, 1979)
Dixon, Heather, 'Steady on Course', *Yachting World* (Annual, 1967), pp. 34–7
Grant, G. and Klinkert, J., *The Ship's Compass* (Routledge, 1952)
Hitchins, H.L. and May, W.E., *From Lodestone to Gyro-Compass* (Hutchinson, 1952)
Lewis, Richard, *The Life-Boat and its Work* (Macmillan, 1874)
May, W.E., *A History of Marine Navigation* (G.T. Foulis, 1973)
Royal Geographical Society, *Navigation through the Ages* (John Murray, 1948)
Warner, Deborah Jean, 'Compasses and Coils: The Instrument Business of Edward S. Ritchie', *Rittenhouse* (1994), vol. 9, no. 1, pp. 1–24

William Thomson's Depth Sounder
Admiralty, *Manual of Seamanship* (HMSO, 1937), vol. i
Bennett, J.A., *The Divided Circle* (Phaidon Christie's, 1987)
Carr, Frank, *The Yachtsman's England* (Seeley Service, 1937)
Carr, Frank, *The Yacht Master's Guide* (Davies, 1940)
Desmond, Kevin, *The Guinness Book of Motorboating Facts and Feats* (Guinness Superlatives, 1979), pp. 220–1
Paasch, Henri, *Illustrated Marine Encyclopedia* (Paasch, 1890)
Russell, Alexander, *Lord Kelvin* (Blackie & Son, 1938)
Thompson, Silvanas, *The Life of Lord Kelvin* (Macmillan, 1910)
Young, A.P., *Lord Kelvin* (British Council, 1948)

Ernst Abbe's Binoculars
Auerbach, Felix, *The Zeiss Works and the Carl Zeiss Foundation in Jena* (Foyle, 1927)
Gause, Hans, 'Ernst Abbe in Memomorian', *Jena Review* (1965), no. 1, pp. 71–5
Gubas, Larry, 'Zeiss Binoculars', *Zeiss Historica* (Spring 1990), pp. 5–14
Reid, William, 'Battling Binoculars', *Zeiss Historica* (1995), vol. 17, no. 2, pp. 3–6
Reid, William, 'The Admirable Barr & Stroud 7 x 50', *Bulletin of the Scientific Instrument Society* (1997), no. 54, pp. 15–20
Reid, William, 'Zeiss and Ross', *Zeiss Historica* (2001), vol. 23, no. 1, pp. 8–15

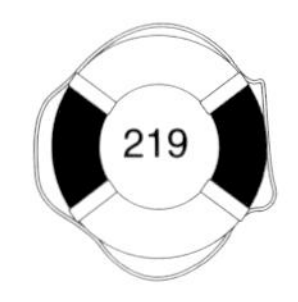

Watson, Fred, *Binoculars, Opera Glasses and Field Glasses* (Shire, 1995)

Zeiss Historica, '100 Years of Carl Zeiss Binoculars' (Spring 1994), pp. 4–6

CHAPTER 3. WEATHER AND SIGNALS

Luke Howard's Cloud Classification

Dictionary of National Biography (OUP, 1921–2), vol. x, pp. 51–2

Hamblyn, Richard, *The Invention of Clouds* (Picador, 2001)

Howard, Luke, *On the Modification of Clouds* (Taylor, 1894)

Watts, Alan, *Instant Weather Forecasting* (Adlard Coles, 2000)

Francis Beaufort and the Wind Scale

Courtney, Nicholas, *Gale Force 10* (Review, 2003)

Dictionary of National Biography (OUP, 1921–2), vol. ii, pp. 39–40

Forrester, Frank, 'How Strong is the Wind?', *Weatherwise* (1986), vol. 39, pp. 147–51

Friendly, Alfred, *Beaufort of the Admiralty* (Hutchinson, 1977)

Fry, H.T., 'The Emergence of the Beaufort Scale', *Mariner's Mirror* (1967), vol. 53, pp. 311–13; and (1968), vol. 54, p. 412

Johnson, Erik, 'On Wind and Water', *Classic Boat* (December 1998), no. 126, pp. 44–7

Frederick Marryat's Signal Flags

Buster, Alan, *Captain Marryat* (University of California Press, 1980)

Dictionary of National Biography (OUP, 1921–2), vol. xii, pp. 1086–8

Ellis, S.M., 'Captain Frederick Marryat', *Bookman* (August 1922), pp. 198–204

Hannay, David, *Life of Frederick Marryat* (Walter Scott, 1889)

Kemp, Peter, *The Oxford Companion to Ships and the Sea* (OUP, 1988)

Marryat, Frederick, *A Code of Signals for the use of Vessels Employed in the Merchant Service* (London, 1841)

Mead, Hilary, 'Flag Signalling for Yachts', *Mariner's Mirror* (1932), vol. 18, pp. 375–81

Woods, David L. (ed.), *Signalling and Communicating at Sea* (Arno, 1980)

William Evans's Navigation Sidelights

Admiralty, *Manual of Seamanship* (HMSO, 1937)

Hayman, Bernard, *Yachting Signalling* (Nautical, 1983)

Hudson, Herbert, 'The Lens of Navigation Lights', *Yachting World* (26 March 1927)

Kemp, Peter, *The Oxford Companion to Ships and the Sea* (OUP, 1988)

Nautical Magazine (May 1845), vol. 3, pp. 725–6

Rettie, Robert, *On the Necessity of Employing One Universal System of Marine Night Signals* (London, 1847)

Senior, W., 'The Beginning of Sidelights', *Mariner's Mirror* (September 1913), pp. 257–64

Thomas, W.R., 'Captain William Evans', *British Chess Magazine* (January 1928)

Philip Columb's Morse Lamp

Columb, Philip, *Naval Warfare* (1891)

Columb, Philip and Bolton, F.J., *Flashing Signals* (Nunn, 1870)

Dictionary of National Biography (OUP, 1917), pp. 207–9

Kemp, Peter, *The Oxford Companion to Ships and the Sea* (OUP, 1988), p. 803

Padfield, Peter, *Aim Straight* (Hodder and Stoughton, 1966)

Robert Fitzroy, Weather Forecaster

Dictionary of National Biography (OUP, 1921–2), vol. vii, pp. 207–9

FitzRoy, Robert, *The Weather Book* (Longman Green, 1862)

Gribbin, John and Mary, *Fitzroy* (Review, 2003)

Hart-Davis, Adam and Bader, Paul, *More Local Heroes* (Sutton, 1998), pp. 32–5

Hughes, Patrick, 'Fitzroy the Forecaster', *Weatherwise* (1988), vol. 41, pp. 200–4

Lewis, Richard, *The Lifeboat and its Work* (Macmillan, 1874)

Mellish, H., *Fitzroy of the Beagle* (Mason & Lipscomb, 1968)

Lee de Forest's Radio Telephone

All Hands, 'Navy's Role in Radio' (June 1957), pp. 44–7

Blanchard, Walter F., 'Radio-Navigation's Transition from Sail to Steam', *Journal of Navigation* (1997), vol. 50, no. 2, pp. 155–71

De Forest, Lee, *Father of Radio* (Wilcox and Follett, 1950)

Johnson, Peter, *The Guinness Book of Yachting Facts and Feats* (Guinness Superlatives, 1975)

Signal, 'The Navy's Role in the Development of Radio' (September–October 1956), pp. 19–24

Sterling, Christopher H., *Encyclopedia of Radio* (Fitzroy Dearborn, 2004), vol. i

CHAPTER 4. SAILS AND RIGGING

Nathaniel Butler's Bermuda Rig

Arnell, Jack, *Sailing in Bermuda* (Royal Hamilton Amateur Dinghy Club, 1982)

Baader, Juan, *The Sailing Yacht* (Adlard Coles, 1965)

Brewington, M.V., *Chesapeake Bay Log Canoes* (The Mariners' Museum, 1937)

Carr Laughton, L.G., 'The Bermuda Rig', *Mariner's Mirror* (1956), vol. 62, pp. 333–5

Chapelle, Howard, *American Sailing Craft* (Bonanza Books, 1936)

Folkard, Henry, *The Sailing Boat* (Edward Stanford, 1901), pp. 392–6

Lefroy, John, *Memorials of the Discovery and Early Settlement of the Bermudas* (London, 1877)

Ratsey, Ernest and de Fontaine, Wade, *Yacht Sails: Their Care and Handling* (Norton, 1948), pp. 37–49

William Gordon's Spinnaker

Field, The (18 August 1866)

Baader, Juan, *The Sailing Yacht* (Adlard Coles, 1965), p. 167

Drummond, Maldwin and McInnes, Robin, *The Story of the Solent* (Cross, 2001)

Heckstall-Smith, B. and du Boulay, E., *The Complete Yachtsman* (Methuen, 1912)

Johnson, Peter, *The Guinness Book of Yachting Facts and Feats* (Guinness Superlatives, 1975), pp. 233–4

Kemp, Dixon, *A Manual of Yacht and Boat Sailing* (Cox, 1913)

Maudslay Society, *Henry Maudslay, 1771–1831* (Maudslay Society, 1949)

Ratsey, Ernest and de Fontaine, Wade, *Yacht Sails: Their Care and Handling* (Norton, 1948), pp. 107–15

Robert Wykeham-Martin's Furling Jib

Cooke, Francis, *Hints, Tips and Gadgets* (Edward Arnold, 1939), pp. 90–1

Fishwick, Mark, 'If I can't make it myself or buy it from Woolworth's', *Classic Boat* (September 1990), pp. 39–44

Fishwick, Mark, 'Rolling Down the Years', *Classic Boat* (October 1990), pp. 35–40

Griffiths, Maurice, *Dream Ships* (Hutchinson's, 1949)

Sven Salén's Genoa

Curry, Manfred, *Yacht Racing: The Aerodynamics of Sails, and Racing Tactics* (Bell, 1928)

Gougeon, Meade and Knoy, Ty, *The Evolution of Modern Sailboat Design* (Winchester, 1973), pp. 103–4

Hoyt, Sherman, *Yankee Yachtsman* (Harrap, 1951)

Loibner, Dieter, *The Folkboat Story* (Sheridan House, 2002)

Ratsey, Ernest and de Fontaine, Wade, *Yacht Sails: Their Care and Handling* (Norton, 1948), pp. 94–6, 103–7, 225–31

Yachting World, 'Racing at Copenhagen' (30 July 1927)

Starling Burgess's Alloy Mast

Laird Clowes, G., *The Story of Sail* (Eyre & Spottiswoode, 1936)

Rogers, Stanley, *Freak Ships* (John Lane, 1936)

Stephens, Olin, *All This and Sailing Too* (Mystic Seaport, 1999)

Vanderbilt, Harold, *Enterprise: The Story of the Defence of the America's Cup in 1930* (Charles Scribner, 1931)

Yachting World, 'Enterprise: The Machine' (12 September 1930)

Yachting World, 'The America's Cup Reviewed' (26 September 1930)

Rex Whinfield's Terylene Sail

Brunnschweiler, David and Hearle, John (eds), *Tomorrow's Ideas and Profits* (The Textile Institute, 1993)

Cowper, Frank, *Yachting and Cruising for Amateurs* (Bazaar, 1921), pp. 30–1

Howard-Williams, Jeremy, *Sails* (Adlard Coles, 1976)

Johnson, Peter, *The Guinness Book of Yachting Facts and Feats* (Guinness Superlatives, 1975), pp. 229–30

Kennedy, Carol, *ICI: The Company that Changed our Lives* (Paul Chapman, 1993)

Worth, Claud, *Yacht Cruising* (Potter, 1910)

Peter Chilvers's Sailboard

Dunnett, Gregg, 'Who Invented Windsurfing?', *Boards* (September 2003), pp. 69–74

Nicholson, Ian, 'Windsurfer Case', *Yachts & Yachting* (1982)

Chapter 5. Keels and Hulls

William Petty's Catamaran

Amateur Yacht Research Society, *American Catamarans* (1957)

Corlett, E., 'Twin Hull Ships', *Transactions of the Royal Instituition of Naval Architects* (October 1969), pp. 401–38

Dictionary of National Biography (OUP, 1921–2), vol. xv, pp. 999–1005

Fitzmaurice, Edmond, *The Life of Sir William Petty* (John Murray, 1895)

Sharp, Lindsay, 'Sir William Petty and Some Aspects of Seventeenth-Century Natural Philosophy', D.Phil. thesis, Wadham College, 1977

Sisk, Hal, 'L'Invention du Catamaran European', *Le Chasse-Marée* (1992), no. 62, pp. 12–27

John Schank's dagger-board

Charnock, John, *An History of Marine Architecture* (Faulder, 1802)

Clark, Arthur, *The History of Yachting, 1600–1815* (Putnam, 1904), p. 103

Folkard, Henry, *The Sailing Boat* (Edward Stanford, 1901)

Herreshoff, Francis, *An Introduction to Yachting* (Sheridan House, 1961)

Kemp, Dixon, *A Manual of Yacht and Boat Sailing* (Cox, 1913)

Knox-Johnston, Robi, *History of Yachting* (Phaidon, 1990)

MacGregor, David, *Merchant Sailing Ships 1775–1815* (Conway, 1985)

Sullivan, Edward (ed.), *Yachting* (Badminton Library, 1985), vol. 1

Patrick Miller's Trimaran

Amateur Yacht Research Society, *Outrigged Craft* (1956)

Amateur Yacht Research Society, *American Catamarans* (1957)

Barton, Humphrey, *Atlantic Adventurers* (Adlard Coles, 1962)

Clarke, Derrick, *The Multihull Primer* (Adlard Coles, 1976)

Clarke, Derrick, *Trimaran Development* (Adlard Coles, 1972)

Dictionary of National Biography (OUP, 1921–2), vol. xiii, pp. 417–20

Gougeon, Meade and Knoy, Ty, *The Evolution of Modern Sailboat Design* (Winchester, 1973)

Miller, Patrick, *The Elevation, Section, Plan and Views of a Triple Vessel* (Edinburgh, 1787)

Smiles, Samuel (ed.), *James Nasmyth Engineer* (John Murray, 1885)

Edward Bentall's Fin Keel

Bentall, Shirley F., *The Charles Bentall Story* (Bentall Group, 1986)

Chatterton, E. Keble, *Sailing Ships* (Sidgwick & Jackson, 1909)

Folkard, Henry, *The Sailing Boat* (Edward Stanford, 1901)

Heckstall-Smith, B. and du Boulay, E., *The Complete Yachtsman* (Methuen, 1912)

Heaton, Peter, *Yachting: A History* (Batsford, 1955)

Herreshoff, Francis, *Captain Nat Herreshoff* (Sheridan House, 1996)

Jackson, Robert, 'A Brief History of E.H. Bentall & Company', *Farm and Horticultural Equipment Collector* (July–August 2000), pp. 8–9

Kemp, Dixon, *Yacht Architecture* (Cox, 1891)

Kemp, Dixon, *A Manual of Yacht and Boat Sailing* (Cox, 1913), pp. 245–8

Kemp, Peter, *The Bentall Story: Commemorating 150 Years Service to Agriculture, 1805–1955* (Maldon, 1955)

Pritchett, Robert T., *Report on the Collection of Yacht Models in the South Kensington Museum* (1899)

Sullivan, Edward (ed.), *Yachting* (Badminton Library, 1985), vol. 1

Thompson, John, *Horse-Drawn Farm Implements* (Fleet, 1979), pt 2

Robin Balfour's Twin Keel

Balfour, Robin, *A Life, a Sail, a Changing Sea* (Beverley Hutton, 1993)

Balfour, Robin A., *Nine Lives in One* (Sheffield Academic, 1980)
Garrett, Alistair and Wilkinson, Trevor, *The Royal Cruising Club 1880–1980* (1980)
Griffiths, Maurice, *Sixty Years a Yacht Designer* (Conway, 1988)
Muncaster, Martin, 'Twin-Keeled Balfour', *Classic Boat* (March 1993), pp. 70–3
Riverdale, Lord, 'Twin Keel Yachts – Development over 45 Years', *Transactions of the Royal Institution of Naval Architects* (July 1968), pp. 327–45
Tucker, Robert, 'Bilge Keels', *Yachting Monthly* (April 1958), pp. 184–7
Yachting World, 'The Bluebirds of Thorne', *Yachting World* (Annual, 1965), pp. 50–5

Uffa Fox's Planing Dinghy
Dixon, June, *Uffa Fox* (Angus & Robertson, 1978)
Dictionary of National Biography (OUP, 1986), pp. 324–5
Du Plessis, Hugo, *Fibreglass Boats* (Adlard Coles, 2002)
Fox, Uffa, *Sailing, Seamanship and Yacht Construction* (Peter Davies, 1934)
Fox, Uffa, *Thoughts on Yachts and Yachting* (Peter Davies, 1938)
Johnson, Peter, 'Uffa Fox – Designer Extraordinaire', *Yachting World* (April 1994), pp. 86–8
Spurr, Daniel, *Heart of Glass* (International Marine, 1999)
Vaughan, T.J., *The International Fourteen Foot Dinghy* (International Fourteen Foot Dinghy Class Association of Great Britain, 1989)
Vaughan, Tom, 'Fourteen Feet Revolution, *Classic Boat* (September 1992), pp. 50–4

John Illingworth, Ocean Racer
Dear, Ian, 'John Illingworth: Man and Myth', *Classic Boat* (November 1999), pp. 32–7
Heaton, Peter, *Yachting: A History* (Batsford, 1955)
Illingworth, John, *Offshore* (Adlard Coles, 1958)
Illingworth, John, *The Malham Story* (Nautical, 1972)
Johnson, Peter, 'John H. Illingworth and *Myth of Malham*', *Yachting World* (February 1994), pp. 50–1
Leather, John, 'The Innovative Mr Giles', *Classic Boat* (June 2001), pp. 34–41
Maynard, Rob, 'Laurent Giles Unlimited', *Classic Boat* (September 1989), pp. 33–7

CHAPTER 6. ENGINES, STEERING AND ANCHORS

John Hawkins's Steering Wheel

Charnock, John, *An History of Marine Architecture* (Faulder, 1802)

Commissioners of Patents for Inventions, *Patents for Inventions: Abridgement of Specifications Relating to Steering* (HMSO, 1875)

Folkard, Henry, *The Sailing Boat* (Edward Stanford, 1901)

Harland, John, 'The Early History of the Steering Wheel', *Mariner's Mirror* (1 February 1972), vol. 58, pp. 41–68

Lewis, M., *The History of the Royal Navy* (Penguin, 1957), pp. 87–92

McGowan, Alan, *The Ship: 4* (HMSO, 1981), p. 16

Paasch, Henri, *Illustrated Marine Encyclopedia* (Paasch, 1890)

Waters, D.W., *The Steering Wheel: Invention, Introduction and Function* (Lisbon, 1985)

Andrew Smith's Wire Rigging

Brett, Edwin, *Notes on Yachts* (Sampson Low, 1869)

Forrestier-Walker, E., *The Wire Rope Industry of Great Britain* (Federation of Wire Rope Manufacturers, 1952)

Kemp, Dixon, *Yacht Architecture* (Cox, 1891)

Martin, J.C., 'The Development of Wire Rope', *International Journal of Maritime History* (June 1992), pp. 101–20

Francis Smith's Screw Propeller

Anon., *On the Introduction and Progress of the Screw Propeller* (Longmans, 1856)

Crabtree, Reginald, *The Luxury Yacht from Steam to Diesel* (Drake, 1974)

Gardiner, Robert (ed.), *The Steam Warship 1815–1905* (Conway Maritime, 1992)

Griffiths, Denis, *Steam at Sea* (Conway Maritime, 1997)

Griffiths, Denis, Lambert, Andrew and Walker, Fred, *Brunel's Ships* (Chatham, 1999)

Lambert, Andrew, 'Responding to the Nineteenth Century: The Royal Navy and the Introduction of the Screw Propeller', *History of Technology* (1999), vol. 21, pp. 1–28

Seaton, Albert, *The Screw Propeller* (Charles Griffin, 1909)

Wilson, Robert, *The Screw Propeller: Who Invented It?* (Thomas Murray, 1880)

Nathaniel Herreshoff and Self-steering

Belcher, Bill, *Wind-Vane Self-Steering* (David & Charles, 1982)

Dufferin, Frederick, *Letters from High Latitudes* (John Murray, 1857)

Ellison, Michael (ed), *Self-Steering* (Amateur Yacht Racing Society, 1999)

Herbert, Tom, *Self-Steering* (Amateur Yacht Racing Society, 1974)

Herreshoff, Francis, *Captain Nat Herreshoff* (Sheridan House, 1996)

Priest, B.H. and Lewis, J.A., *Model Racing Yachts* (Model Aeronautical Press, 1966)

Pritchett, Robert, *et al.*, *Yachting* (Longman, 1894)

Rudolf Diesel's Boat Engine

Cummins, C. Lyle, *Diesel's Engine, Vol. 1, From Conception to 1918* (Carnot, 1993)

Del Vecchio, Mike, 'Remembering Rudolf Diesel: A Hundred Years Ago', *Railfan and Railroad* (1996), vol. 15, no. 5, pp. 38–9

Desmond, Kevin, *The Guinness Book of Motorboating Facts and Feats* (Guinness Superlatives, 1979), pp. 21–2

Evans, Arthur, *The History of the Oil Engine* (Sampson Low, 1932)

Fitzgerald, R., 'Herbert Akroyd Stuart and Diesel's Engine', *Model Engineer* (December 1998), vol. 181, no. 4082, pp. 671–3

Job, Barry, 'The Oil Engine Rivals', *Stationary Engine Magazine* (April 1997), no. 278, pp. 18–20

Smith, Edgar, *A Short History of Naval and Marine Engineering* (CUP, 1938)

Wittmann, Josef, 'Rudolf Diesel: The Man and his 100-Year-Old Engine', *Interdisciplinary Science Review* (September 1994), vol. 19, no. 3, pp. 201–10

Worth, Claud, *Yacht Cruising* (Potter, 1910)

Cameron Waterman's Outboard Motor

Cooke, Francis, *Hints, Tips and Gadgets* (Edward Arnold, 1939)

Desmond, Kevin, *The Guinness Book of Motorboating Facts and Feats* (Guinness Superlatives, 1979), pp. 188–9, 196

Dulken, Stephen van, *Inventing the 20th Century* (British Library, 2000)

Du Boulay, Ernest, *A Textbook on Marine Engines* (The Yachtsman, 1902)
Rodengen, Jeffrey, *Evinrude, Johnson and the Legend of OMC* (Write Stuff Enterprises, 1993)
Schult, Joachim, *Curious Boating Inventions* (Paul Elek, 1974)
Voss, John, *Venturesome Voyages of Capt. Voss* (Geiger & Gilbert, 1913)
Warring, R.H., *Outboard Motors and Overdrives* (Pitman, 1966)

Geoffrey Taylor's CQR Anchor
Cooke, Francis, *Hints, Tips and Gadgets* (Edward Arnold, 1939)
Curryer, Betty Nelson, *Anchors: An Illustrated History* (Chatham, 1999)
Fagan, Brian, *Staying Put* (Fernhurst, 2002)
Garrett, Alistair and Wilkinson, Trevor, *The Royal Cruising Club 1880–1980* (1980), pp. 115–17

CHAPTER 7. EMERGENCIES
Lionel Lukin's Lifeboat
Bennett, Susan, 'Lifeboat', *Journal of the Royal Society of Arts* (May 1996), p. 56
Cunliffe, Tom, 'The Old Original', *Yachting Monthly* (September 1990), pp. 116–17
Dawson, A., *Britain's Life-Boats* (Hodder & Stoughton, 1923)
Fry, Eric, *Life-Boat Design and Development* (David & Charles, 1975)
Lamb, John Cameron, 'The Lifeboat and its Work', *Journal of the Royal Society of Arts* (18 February 1910), pp. 354–9
Leach, Nicholas, *Lifeboats* (Shire, 1998)
Lewis, Richard, *The Life-Boat and its Work* (Macmillan, 1874)
Osler, Adrian, *Mr Greathead's Lifeboats* (Tyne & Wear Museums Service, 1990)

George Manby's fire extinguisher
Desmond, Kevin, *The Guinness Book of Motorboating Facts and Feats* (Guinness Superlatives, 1979), p. 211
Ffiske, William H., *A Short History of Early Life-Saving Apparatus* (Ffiske, 1925)
Hart-Davis, Adam and Bader, Paul, *More Local Heroes* (Sutton, 1998)
Manby, George, *The Extinction and Prevention of Destructive Fires* (London, 1830)

Manby, George, *A Brief and Faithful Historic Sketch* (Manby, 1852)
Mechanic's Magazine (19 November 1836)
Walthew, Kenneth, *From Rock and Tempest* (Geoffrey Bles, 1971)
Wright, Brian, *Firefighting Equipment* (Shire, 1989)

Peter Halkett's Inflatable Cloak-boat
Blandford, Percy, *An Illustrated History of Small Boats* (Spurbooks, 1974)
Desmond, Kevin, *The Guinness Book of Motorboating Facts and Feats* (Guinness Superlatives, 1979), pp. 167–9
Halkett, Peter, *Boat-Cloak or Cloak-Boat* (Wall & Hiscoke, 1848)
Hancock, Thomas, *Personal Narrative* (Hancock, 1920)
Mariner's Mirror, 'The Halkett Boat and Other Portable Boats' (1958), vol. 44, pp. 154–8
Worcester, G., 'The Inflated Skin Rafts of the Huang Ho', *Mariner's Mirror* (1957), vol. 43, pp. 73–4
Zodiac Group, *A Century of Air and Water* (Zodiac, 1996)

Edward Berthon's Folding Dinghy
Berthon, Edward, *A Retrospect of Eight Decades* (George Bell, 1899)
Cowper, Frank, *Boat-Sailing for Amateurs* (Exchange & Mart, 1922)
Dictionary of National Biography (OUP, 1921–2), vol. xxii, pp. 184–5
Kemp, Dixon, *A Manual of Yacht and Boat Sailing* (1891)
Sullivan, Edward (ed.), *Yachting* (Badminton Library, 1985), vol. 1
Worth, Claud, *Yacht Cruising* (Potter, 1910)

John Ward's Lifejacket and the First Survival Suits
Clemans, Martyn, 'A Matter of Survival', *Mariner's Mirror* (1965), vol. 51, pp. 253–4
Desmond, Kevin, *The Guinness Book of Motorboating Facts and Feats* (Guinness Superlatives, 1979)
Lewis, Richard, *The Life-Boat and its Work* (Macmillan, 1874)
The Yachtsman (8 July 1897), p. 25

Martha Coston's Signalling Flare

Barnett, J.P., *The Lifesaving Guns of David Lyle* (South Bend Replicas, 1976)

Coston, Martha J., *A Signal Success: The Work and Travels of Mrs. Martha J. Coston* (Lippincott, 1886)

Pilato, Denise E., *The Retrieval of a Legacy: Nineteenth-Century American Women Inventors* (Praeger, 2000)

Pilato, Denise E., 'Martha Coston: A Woman, a War, and a Signal to the World', *International Journal of Naval History* (April 2002)

Sherwood, Martyn, *Coston Gun* (Geoffrey Bles, 1946)

Vare, Ethie Ann and Ptacek, Greg, *Patently Female* (John Wiley, 2002)

Guglielmo Marconi's S-O-S

Kemp, Peter, *The Oxford Companion to Ships and the Sea* (OUP, 1988)

Reynolds, Pam, *Guglielmo Marconi* (Marconi Company, 1974)

Thompson, J.C., 'The Global Maritime Distress Safety System', *Telecommunication Journal* (1992), vol. 59, no. 1

Wedlake, G.E.C., *SOS: The Story of Radio-Communication* (David & Charles, 1977)

Daniel Mowrey's Seasickness Remedy

Bessemer, Henry, *An Autobiography* (Offices of Engineering, 1905)

Dutton, Thomas, *Sea-Sickness* (Bailliere, 1891)

Illingworth, John, *Offshore* (Adlard Coles, 1958)

Keevil, J.J., *Medicine and the Navy* (Livingstone, 1957), vol. i

Mowrey, Daniel and Clayson, Dennis, 'Motion Sickness, Ginger, and Psychophysics', *The Lancet* (20 March 1982), pp. 655–7

Pingree, B., *Journal of Royal Naval Medical Service* (1989), vol. 75, pp. 75–84

Index

Index

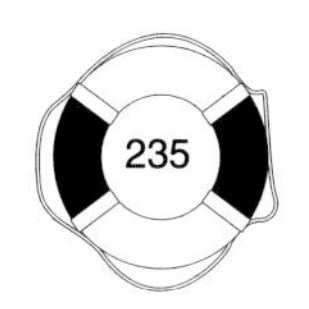

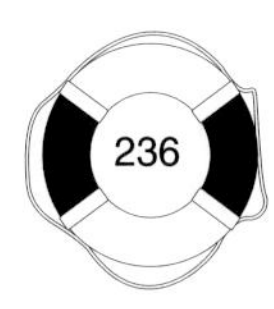

237

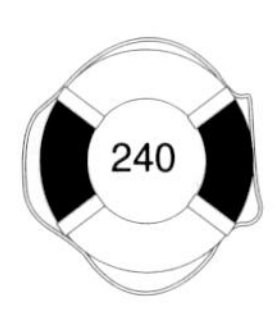

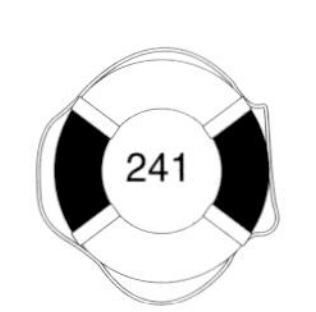